Visualizing Software

A Graphical Notation for Analysis, Design, and Discussion

William S. Bennett
CAE-Link Corporation
Binghamton, New York

Marcel Dekker, Inc.

New York • Basel • Hong Kong

Library of Congress Cataloging-in-Publication Data

Bennett, William S.
  Visualizing software : a graphical notation for analysis, design, and discussion / William S. Bennett.
     p.   cm.
  Includes bibliographical references and index.
  ISBN 0-8247-8714-5
  1. Computer software. 2. Computer graphics.
QA76.754.B45 1992
005.3--dc20   92-13677
                                                              CIP

This book is printed on acid-free paper.

Copyright © 1992 by MARCEL DEKKER, INC. All Rights Reserved.

Neither this book nor any part may be reproduced or transmitted in any form or by any means, electronic or mechanical, including photocopying, microfilming, and recording, or by any information storage and retrieval system, without permission in writing from the publisher.

MARCEL DEKKER, INC.
270 Madison Avenue, New York, New York 10016

Current printing (last digit):
10 9 8 7 6 5 4 3 2

Preface

This book introduces a graphical notation for software, which is intended to improve the understanding of the software not only by its developers but by anyone else who has to understand how it works. The claim we make is that the new notation improves the mental simulation of software — making it easier to run through the operation of the software in the mind, and giving greater assurance that the design is correct.

Because of that, we also make the claim that the notation will improve the analysis and design of the software and make it easier to communicate its operation to others in discussions and design reviews.

The problem in the design of software notations, whether alphanumeric or graphical, is that there are so many different aspects of software which its developers must understand. It is difficult to choose which ones to put into a notation; over a period of more than forty years, several sets of aspects have been tried. One of the earliest focused on control flow in executable processes, followed in the 1970's by a shift to the data flow between and among processes. More recently there has been an emphasis on partitioning, first by function and now by "objects" — stressing the protection of data through its encapsulation with the processes that manipulate it, and stressing a strict correspondence between these encapsulated objects and the real-world entities they control or represent.

Through all of this, however, the problem of understanding the operation of the software, and of making sure that it does the right thing, has remained the most difficult to solve. Data flow analysis and object encapsulation have helped to break the software into manageable pieces, but the individual pieces must still be understood.

What we have done here is to select a new subset of the many aspects of software — a subset we believe to be important to mental modeling of the behavior of the software. These aspects are then placed on the same diagram.

When all the parts relevant to mental simulation can be seen at once, with all their interactions, then almost nothing need be remembered from some other document; relieved of that burden, our minds can explore the diagram and recreate all the actions of the software. In turn, that enables us to see whether the software is doing what we want it to do, to spot errors and see how to fix them, and to invent new and better ways for the software to work.

The parts of the software that we believe are relevant to mental simulation are:

- Processes — that is, executable code;

- Data, either in memory or on more permanent storage;

- The flow of data among processes;

- The flow of control, through sequential and parallel paths;

- Decisions;

- The action of data on decisions;

- The action of decisions on control flow; and

- The action of control flow on processes.

The graphical notation introduced in this book contains all of these parts present in one kind of diagram, in a direct and simple way that is intuitively understood. A different kind of icon is used for each part, and their interconnections capture all of the interactions.

The notation permits the software to be organized hierarchically: processes can be expanded to see what is in them, to any level; decisions can be expanded in much the same way; and aggregations of data can be shown splitting, level by level, into their components, or being built up from them.

Preface

Then, when the software is expressed this way, simple enclosures around various groups of icons can express states and state transitions, software objects and their operations, the division of software into encapsulated or protected entities, and the allocation of software to different physical processing units. These enclosures can be extended to many other useful segmentations.

This book explains the notation for those who want to learn how to use it. We have tried to make it as clear and understandable an exposition of the notation as we can. Although there is some discussion of other notations, the book is not a complete comparative analysis of all software notations, either alphanumeric or graphical. Many examples are given, including one fairly complete example, but all are necessarily small. We could not, in the space of this book, take a large project through all the phases of software design and production, showing all its diagrams. However, we hope that the examples given will serve to illustrate not only how to use the notation but its basic advantages.

Similarly, although there has been some considerable work reported in the literature on mental modeling and mental simulation in general, this book is not a survey of that work or of the way in which it pertains to mental simulation of software. A scientific study of mental modeling with this new notation, as against other notations, has not yet been performed.

The notation is in use, for different purposes, on several programs (both contractual and internal) at the CAE-Link Corporation, a manufacturer of flight training simulators. It has been selected for use as a basic documentation method for algorithms (particularly those associated with software objects) on a number of future programs. Besides its use for algorithms, it has been used to express operational concepts and as a way to explain software in various kinds of meetings. It is beginning to be used as a way to coordinate software development among companies on a multi-company project. In all these, it has been found to be especially useful where the software is complex and difficult to understand.

It is important to understand as well that this is not a software development method — it is a notation, and it can be used with any method that fits your organization and your application. Even when a certain method is being applied, software usually gets developed in a convoluted way; but a good notation will make sure that the product, at any stage including final, is logically consistent and fully represented.

In general, notations such as this one require some form of computerization in order to be fully usable. Programmable automated drawing tools have been adapted for this notation, for example, the Sylva system offered by Cadware, Inc., on which an early version of the notation is commercially available, and the Design OA tool offered by Meta Software, Inc., which has been adapted to the notation by work at CAE-Link Corporation and is being used within the company. Those who wish further information about these tools are invited to write to the author at:

CAE-Link Corporation
P. O. Box 1237
Binghamton, New York 13902-1237

Alternatively, the notation can be used with widely available drawing tools such as the MacDraw II package offered for the Macintosh line of personal computers (which was used for the illustrations in this book), although without some of the features offered by the Sylva and Design OA tools.

The use of the notation itself is unrestricted. It is offered as a clear and easily understood way to document software designs and to communicate those designs to others.

The author would like to acknowledge all of his colleagues at CAE-Link Corporation who offered suggestions on the content of the book, especially Dr. Jeffrey Kleinwaks, Robert Epps, and William Wessale.

William S. Bennett

Contents

Preface iii

Introduction: Seeing Software with the Mind's Eye 1

Basics of the Notation 17
Processes 17
Data 17
Data Flow into and out of Processes... 18
Control Flow 19
Action of Control Flow on Processes 19
Decision Blocks 21
Action of Data on Decision Blocks 21
Decision Blocks with One Output..23
The Structured Constructs in this Notation 25
Parallel Control Flow Paths and Symbols 28
Independence of Parallel Paths 29

Interconnected and Hierarchical Diagrams 31
"Flat" Connections via Page Connectors 31
A Better Way for Many Tasks: Hierarchical Interconnection 32
Hierarchical Designing and Debugging 33
Structuring Your Diagrams 34
Hierarchies of Processes 36
Data Flow Between Higher and Lower Diagrams 37
Hierarchies of Decisions 38
Hierarchies of Data 39

A Detailed Example: A Digital Wristwatch 41
  1.0 Wristwatch 43
  1.1 Time Display and Control 45
  1.1.1 Normal Time Display / Control 49
  1.1.1.1 Normal Time Set 51
  1.1.1.2 Reminder Control 53
  1.1.2 Alarm Display / Control 53
  1.1.2.1 Chime Set 56
  1.1.2.2 Alarm Set 57
  1.1.3 Dual Time Display / Control 57
  1.1.3.1 Dual Time Set 57
  1.1.4 Timer Time Display / Control 61

Contents

1.1.4.1 Set Timer Time Setting 61
1.1.5 Chrono Time Display / Control 61
1.2 Ring Alarms 65
1.2.1 Alarms? 67
1.3 Measure Time 69
1.3.1 Normal Time Checks 71
1.4 Light Control 73

Separation Through Enclosures 75
Enclosures 75
States and State Transitions 76
Decision Blocks and State Transitions 76
Active and Passive States 77
A Flight Training Simulator 78
Physical CPU Enclosures 80

Software Objects 81
Object Enclosures 81
Operations of Objects 82
Shared Data Structures 82
Software Objects in Industrial Software: Control of Software Changes 87
Software Objects in Flight Training Simulation 88

Communicating Concurrent Processes 97
An Automobile Cruise Control 97
The Cruise Control in Parallel Form 100
Communication Using Data Flags 104
Hiding the Details with a Hierarchy 104

Interrupts 111
Enclosing Interruptible Software 112
Suspend and Resume Control 113
Controlling the Interrupt on a Parallel Path 114
The Interrupt as a Data Signal 114

Using the Notation for Software Requirements 117
Functional Requirements 118
Explosion of Data Requirements 118
Annotation with Data Attributes 120
Annotation with Timing Requirements 121
Required Algorithms 122

Rules of the Notation 125

An Agricultural Inspection Device 137

A Measurement System 149

Contents

A Simple Camera Control System 155

Clarifying Your Drawings:
   The "Data Bus" Solution 167

Choosing a CASE Tool 175

A Closing Note 188

References 189

Index 190

Introduction: Seeing Software With the Mind's Eye

There must be, by now, more than a hundred notations for software that have had some sort of publication, and perhaps well over a thousand if various methods developed inside organizations and universities are counted. Why so many?

Early in the development of computers, in the late 1940's and up through the 1960's, it was recognized that the operation of the software — its behavior under various circumstances — was the most important aspect. The wholly alphanumeric (often machine-level) code did not allow the developers to visualize the operation of the software. Control flow was the newest and most exotic aspect of computers, and the most difficult to visualize. The earliest flow charts were an attempt to draw a picture of these processes, and were designed to show just the executable processes and the control flow (and decisions) that caused the processes to be executed.

But as software systems expanded to millions of lines of code, and had to be broken up to be worked on by hundreds or even thousands of people, developers recognized that the interfaces among these many pieces had to be worked out before the code could be written. So, in the 1970's, it became popular to draw diagrams that stressed the interfaces — that is, the data that flowed between processes [2, 3, 10]. Control flow was believed to be of secondary importance, often because business data processing applications had simple or regular patterns of control flow. Later, in the 1980's, special auxiliary diagrams (or alphanumeric documents) containing control flow were reintroduced for "real-time" applications — not so much because rapid response was needed, but because the applications were complex enough to make their behavior obscure [6, 7].

Later in the 1980's, it was recognized that if the software were broken up or partitioned so that each part corresponded to some real-world entity, and had within it everything — data and operations — that pertained to that entity, then when the inevitable changes to those entities came along, each change would affect only one part of the software. Furthermore, these entities were related in classes — you could distill out both data and operations that applied to whole classes, and so not have to re-invent that software every time you applied it to a specific instance. So was born the idea of a software "object," in which the data and operations relating to a certain real-world entity or class would be encapsulated together and kept separate from the rest of the software. And with the software objects came diagrams which stressed this partitioning and its advantages [9, 11,12].

In all these shifts among both alphanumeric and graphical methods of visualizing software, however, the original concept of visualizing the *behavior* of the software has somehow been downgraded. Developers have found themselves wishing that there could be a way to see, in their mind's eyes, how the software will really work. So while diagrams that showed the partitioning of large software systems, or showed the interfaces among its pieces, were clearly of value, still a way was needed to visualize behavior. And—we contend — it is mental simulation that is needed to provide that visualization [4, 5].

Of the methods that are available to provide the visualization, the main categories are alphanumeric and graphical. To some extent the choice between the two is personal and based on the individual's ability to understand in words or in pictures. Design languages based on the use of pseudo-code have been used extensively by those who lean toward words.

We would contend, however, that nearly every human software developer, or manager or customer, can benefit from a clear graphical design representation. Humans, as primates, think naturally with their eyes and with the mind's eye — an internal visual representation. The long history of the use of maps and drawings testifies to that. Even those whose primary orientation is toward alphanumeric software representations use maps and diagrams in many other aspects of their lives, and, we think, could use and appreciate an effective graphical software representation.

Introduction: Seeing Software with the Mind's Eye

Of the graphical notations available that do try to show the operation of the software and therefore make mental simulation possible, though, most have one problem in common: not all the aspects of software that are necessary are modelled graphically within the notation. If a software developer -- or the developer's manager, peer reviewer, or customer -- must try to determine whether the software is behaving properly by inspecting the diagram, the missing aspects must either be borne by that person's memory, or must be supplied by other kinds of documents — mini-specifications, pseudo-code or "design language" statements, state transition networks, and other formats — perhaps referenced in the notation.

When the developer tries to simulate the operation of the software mentally with a notation that does not represent enough aspects of the software, or that puts various aspects on different documents, then a burden is placed on his or her memory that severely constrains that mental simulation, and therefore the developer's ability to understand the action of the software [1, 4, 5].

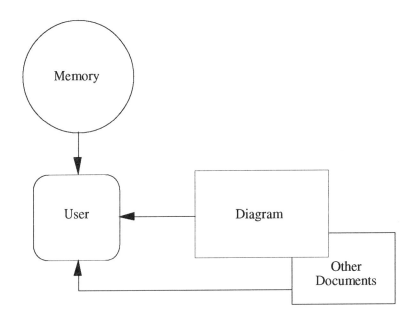

We can show that this is intuitively correct. Take, for example, a road map. Suppose that we had a road map that was presented in two parts: first, a page that had just the dots that represent the towns, say with the town names printed next to them; then, a second page that had just the lines representing the roads, but not the towns.

Now suppose that you want to plan a trip from one town to another, situated some distance away. To plan the trip you must look first on one page to find the town you're starting from, then look at the other page and try to guess which road goes through it.

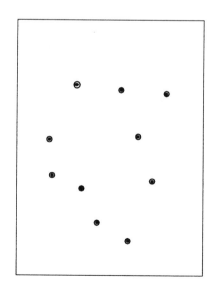
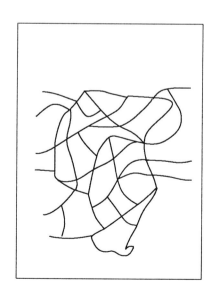

Introduction: Seeing Software with the Mind's Eye

The problem is that the interactions of the roads and towns — that is, the places where a road goes through a town — can't be determined from this divided map. To solve the problem we have to combine the two pages into one diagram, so we can see everything at once.

We wouldn't think of drawing a map with the towns on one page and the roads on another, though we do sometimes draw maps of the same area which show different features. We might have stands of maple trees highlighted on one, and bicycle factories on another. We do this because we don't believe we need to see those features together.

But when we believe it's important to see two features on one map, like roads and towns or like factories and railroads, we put them together, because we want to see their interaction. In a similar way, we should put together on one software diagram those features of a software design that interact.

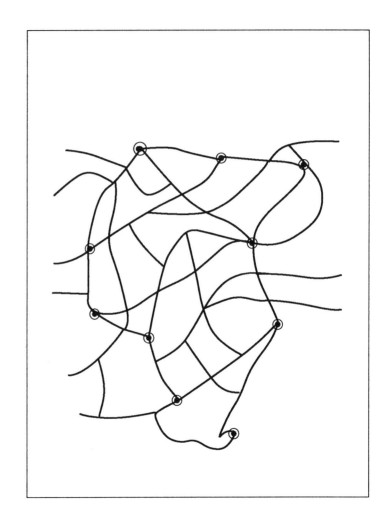

What, then, are these aspects of the software you must consider as essential to this simulation?

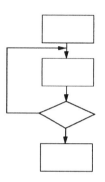

In the 1960's, we thought that control flow was the most important aspect of software; with the "spaghetti coding" that was prevalent then, control flow probably was the most difficult aspect to visualize. So we used the flow chart as the main method of graphical documentation.

Then, in the 1970's (and particularly in business applications with relatively simple processes but very complex disk files), it was felt that the data flow into and out of processes, and into and out of data stores, was the most important aspect. So the data flow diagram became the most important graphic [2, 3, 10].

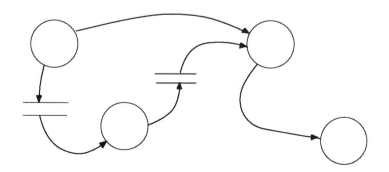

Introduction: Seeing Software with the Mind's Eye

This was sometimes supplemented, though usually not on the same diagram, with a representation of the hierarchical structure of the various modules of the software.

In the 1980's, and especially for applications where rapid response was necessary, we began to feel that control flow needed to be reintroduced and somehow integrated with data flow. But the designers of most notations were still reluctant to return to the flow charts of the 1960's, and so control flow was represented largely just as a line indicating that a process was being started up. The logic that led to that startup was buried in off-diagram documents, typically either pages of pseudo-code or as a separate state transition diagram [6, 7].

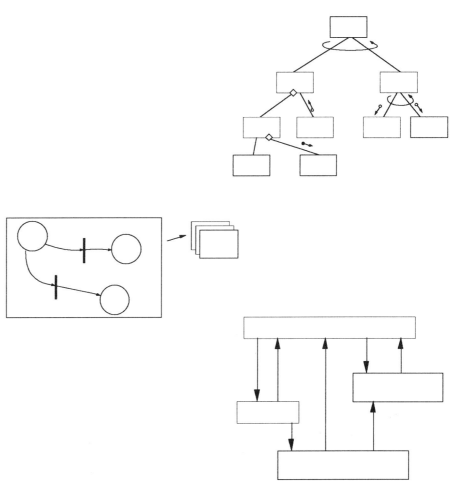

Today, in the 1990's, we are beginning to feel that we really need a fully integrated approach that shows us all these aspects (and more) at once, regardless of the application domain: processes, data, data flow, control flow, decisions, parallel operation, and all their interactions. We used to think that business data processing didn't need this kind of attention; but that was before most modern business applications became so complex, so interrelated, and so interactive both with the user and with each other. The distinction between "business" applications and "real-time" applications has become blurred.

The trouble is that it is difficult to think of a graphical strategem that will integrate data flow, control flow, decisions, state transitions, processes, and all their interactions so that the user can really simulate the action of the software mentally. The notation described in this book is built around the discovery of a graphical means that will integrate those aspects.

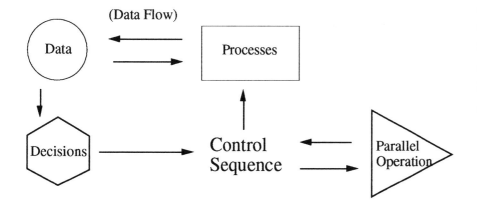

Introduction: Seeing Software with the Mind's Eye

Our notation shows control flow as a dotted arrow, with right-angle stubs threaded along it that point to processes to be done in the order of the threading. This gets the control flow separated so that it doesn't interfere with the arrows showing data flow. Decision blocks are integrated into the control flow network, and connected to the data that drives them. Out of the decision blocks, alternate control flow paths appear (very much like the old 1960's flow charts). Later, in the chapter "Basics of the Notation," we'll cover in detail how this is done.

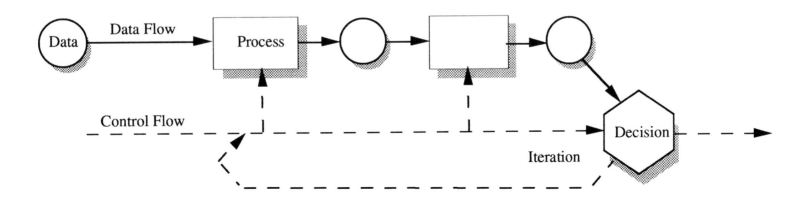

In addition, symbols are provided where parallel control paths are spawned, or rendezvous with one another.

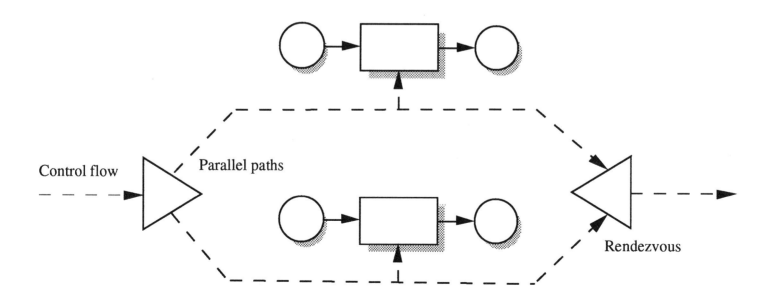

Returning to our map analogy, suppose now that there are some main roads and many secondary roads; and some large towns and many smaller villages. Typically we will have a map which shows the whole area but with just the larger towns and main roads; then we will have several other maps, each of a small part of the area, showing all the towns and all the secondary roads for that part of the territory. It's often helpful to have the maps broken apart in a hierarchy like that — but in each case both roads and towns, and their interactions, are shown on the map.

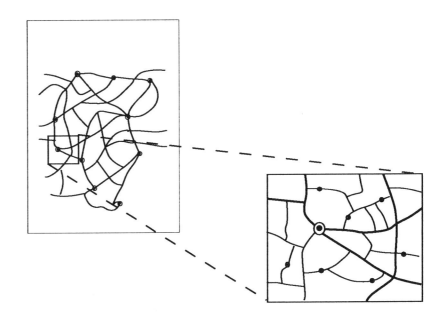

We can use this same principle in dividing up software diagrams. Most of the time, it's impractical to show a large piece of software as one single diagram. Some means of breaking large software systems into many diagrams, like the hierarchical division of the road maps, is essential. When we say that software diagrams should show "all the aspects" of the software, we don't mean that the whole program has to be on one diagram. We mean that within each small piece of the software, it's necessary to show all the aspects that will enable the user to simulate the action of that small piece mentally.

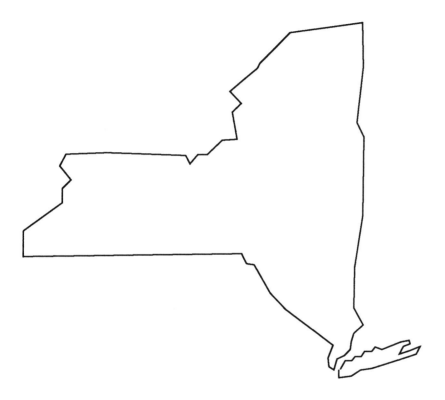

And, once in a while, we find it useful to have a huge road map that has all the towns and all the roads, for a large area. I tend to like those kinds of maps -- I'm not intimidated by the large number of towns and roads, and I like the idea that I can zero in on a small piece of the map and find all the small places I'm looking for.

There will be times when a large software diagram will be useful in the same way; so it's good to have a notation system that can support large diagrams when you need them, yet help you make sense of them. With millions of lines of software, however, even the largest diagrams must be organized hierarchically, so any diagramming system must be able to do that.

Introduction: Seeing Software with the Mind's Eye

So, in our notation, means are provided to partition the diagrams in various ways, first of all in the same hierarchical manner used on the road maps and in many other notations, in which individual processes are exploded into whole diagrams of their own.

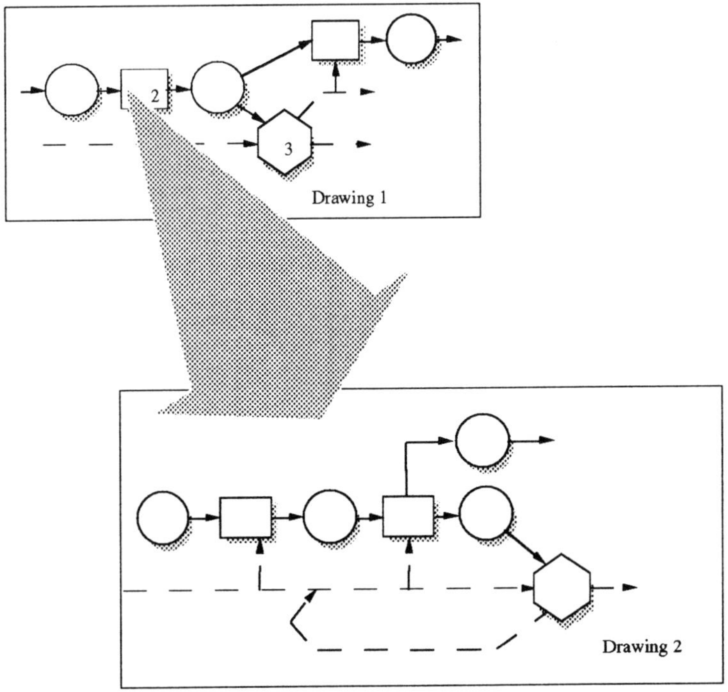

Drawing 1

Drawing 2

In addition, however, because all the relevant aspects are shown, when the appropriate boxes are drawn around regions on the diagram, these diagrams (with no other changes) become detailed state transition diagrams, software object diagrams, physical CPU partitioning diagrams, and any other partitioning that is important to the application at hand.

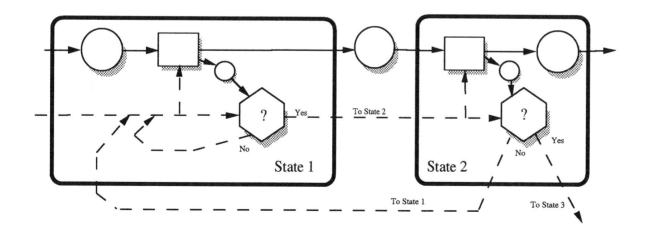

Advantages of the Notation

The advantages of the notation (some of which are, of course, shared by other notations), are:

It is simple, understandable, and requires no lengthy learning curve.

It allows the critical aspects of a piece of software to be shown together on one diagram, so that their interactions are evident. Because of that,

It facilitates mental simulation of the software, and so:
> it fosters correct software designs; and
> it facilitates discussion of software in groups.

It is able to express the hierarchical structure of processes, decisions, and data. Because of that,

It is extendable to large aggregations of software;

It facilitates the division of the software among developers.

It is able to express parallel processing as well as software states, software objects, and many other useful partitionings,

> without losing the mental simulation ability, and

> without losing the full notation.

Since it is only a notation, it can be used with many different groupings of software development work steps, forming "methodologies" which are tailored to specific kinds of software development.

It can be used with most programmable graphical CASE tools, and data sets are available adapting the notation for two such tools.

Basics of the Notation

Processes

A "process," in this book, is an action to be taken by the computer on some data. We represent it as a rectangular block.

This box represents a process

Data

"Data" means any kind of data that can be interpreted by a computer. It can exist in the computer's main memory or on some form of secondary storage such as a disk or tape, or it could be data that is being entered on a keyboard. We represent any of these forms of data as a circle. Data can be anything from numbers to alphanumeric characters to groups of binary bits to whole documents, and can even include one-bit signals such as interrupts (which in some notations are considered "control" signals).

This circle represents data

Data Flow into and out of Processes

We represent data flow into and out of processes as arrows made from solid lines (rather than dotted lines, which we use for control flow). Any data coming into a process block is shown arriving by one of these arrows from a data circle that is external to the block. In general, all processes must generate data (there being no other reason for their existence), and it would be a very rare process which did not require any data as input. This is reflected in a rule that process boxes must show inputs from data circles and outputs to data circles. This much of the notation is, of course, equivalent to a data flow diagram.

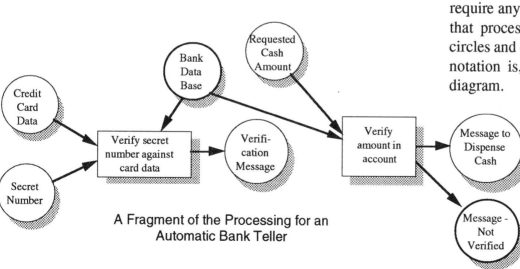

A Fragment of the Processing for an Automatic Bank Teller

Control Flow

We represent the normal flow of control — that is, the sequence of execution of the various process blocks — using a dotted arrow, with control understood to be flowing in the direction of the arrowhead. However, we don't show these arrows as flowing directly into processes, as has been done on flow charts for many years. Instead, we do this:

Action of Control Flow on Processes

When we want to show that the flow of control has been given to a process, we use a short dotted arrow — which we call a stub — coming off the main dotted arrow at right angles and entering the process block. When the control flow coming along the main arrow encounters a stub, this means "execute that process." You can think of the flow of control as coming along the main dotted arrow, going up the stub into the process block, executing it, then coming back out of the process block, down the stub, and continuing along the main arrow.

Control flow moves along this dotted arrow in the direction of the arrowhead

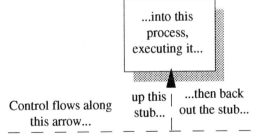

You might wonder why we used this convention instead of just allowing the main dotted arrows to go directly into the process blocks. It has a number of advantages. First, it keeps the control flow structure below, and separated from, the data flow into and out of the process blocks, and therefore makes it possible to appreciate the control flow and data flow separately. At the same time it preserves the interactions between control flow and processes, and between processes and data, and keeps them on the same diagram. And, as we will see below, it makes it easier to appreciate the action of data on control flow, which is difficult to see on other kinds of diagrams.

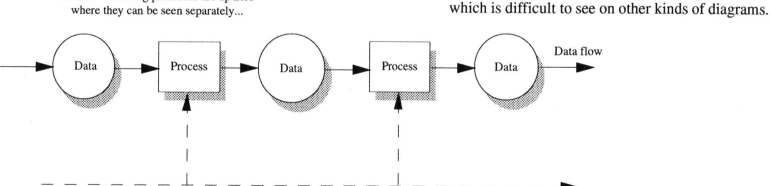

Data flows among processes are up here where they can be seen separately...

While control flow is down here ... still connected, still visible on the same diagram, but clearly separated.

Decision Blocks

We use a hexagon to indicate a decision. A main dotted control flow arrow runs directly into a decision block; then two or more main dotted arrows emerge from the decision, just as was done with diamond decision blocks on flow charts. Usually a question is shown in the decision block, and the answer or condition corresponding to each emerging control flow arrow is marked next to the arrow. It's understood that control flow will continue on one (and only one) of these emerging arrows.

Action of Data on Decision Blocks

Because decision blocks require data with which to make the decision, every hexagonal decision block must be shown as receiving data flows from one or more data circles. However, decision blocks do not emit data, and so data flows may not be shown coming from them.

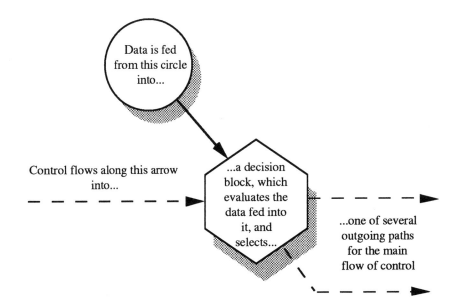

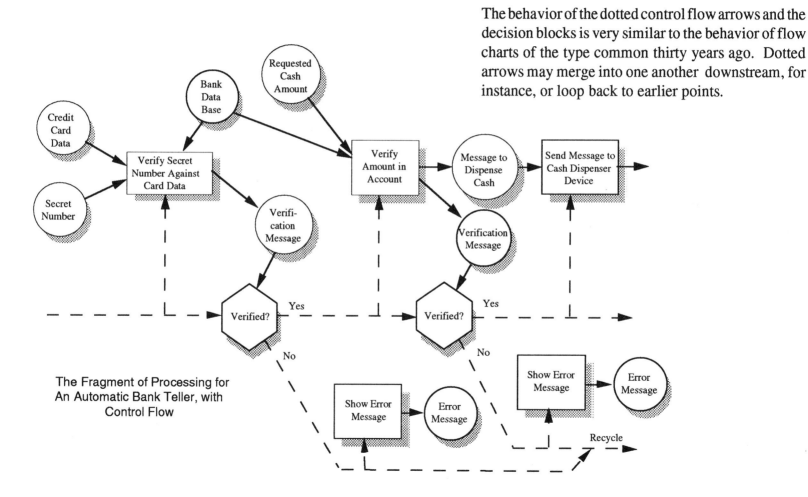

The behavior of the dotted control flow arrows and the decision blocks is very similar to the behavior of flow charts of the type common thirty years ago. Dotted arrows may merge into one another downstream, for instance, or loop back to earlier points.

The Fragment of Processing for An Automatic Bank Teller, with Control Flow

Decision Blocks with One Output

Sometimes the control path will "sit and wait" until some signal arrives. This situation is shown in one of two ways, depending on whether the waiting is accomplished by looping through repeated instructions, or is accomplished through a special circuit like an interrupt chip.

If it's done with repeated instructions, then it's shown with a control path which loops out of the decision block and around back into it. Of course, the decision block must be driven by the signal for which it is waiting, and when that signal is received, control passes out the other dotted control path.

On the other hand, if it's done with a special circuit, or perhaps with the "SUSPEND" statment found in some languages, then it's shown as a decision block with just one control path output. The meaning is clarified by using the words "wait for" in the legend.

If some instructions are being executed repeatedly in a loop, testing whether the signal has arrived, then the loop is shown

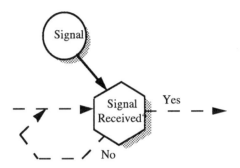

If there's no real looping, the loop isn't shown

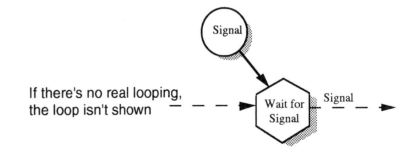

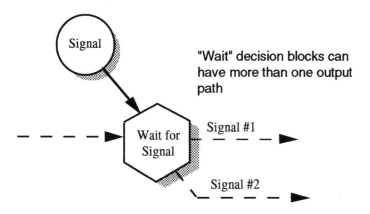

Sometimes, of course, the external signal you're waiting for may be one which selects one of several control paths. If there's no looping during the waiting, the decision block shows just the various output paths, and the words "wait for" appear in the block.

The Structured Constructs in This Notation

The standard structured constructs, "IF-THEN-ELSE", "DO-WHILE," and "DO-UNTIL" are readily diagrammed in this notation. Here, in the IF-THEN-ELSE, the flow of control follows one or the other of the paths. Then the paths merge back together. This is a simple merge; it means simply that control flow follows the dotted arrows. It's not to be confused with the merging of parallel control paths, which are explained later in this chapter.

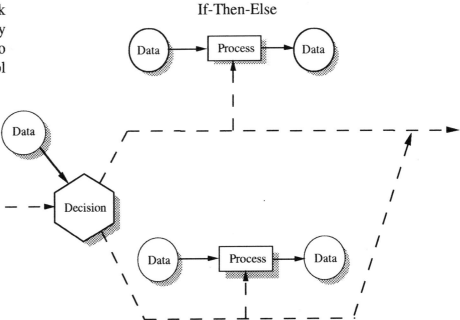

If-Then-Else

In the DO-WHILE, the decision block is at the beginning of the loop and is driven by data developed within the loop.

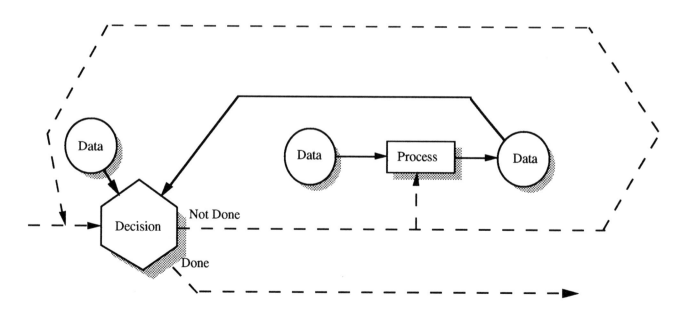

Do While

Basics of the Notation

In the DO-UNTIL, on the other hand, the decision block is at the end of the loop.

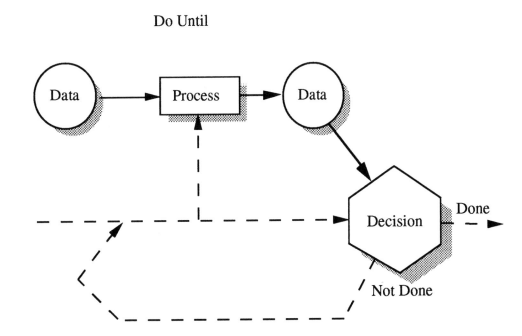

Do Until

Parallel Control Flow Paths and Symbols

At times the flow of control may split into two or more parallel flows, which are then going on simultaneously. We represent this with triangles. The main dotted arrow may run directly into a right-facing triangle, and two or more main dotted arrows, each representing a parallel path, will emerge from it. These flows may recombine at a left-facing triangle. When that happens, it is understood that if one main dotted arrow reaches the triangle first, it will wait until the others have all finished executing their processes and have reached the triangle before control passes beyond the triangle.

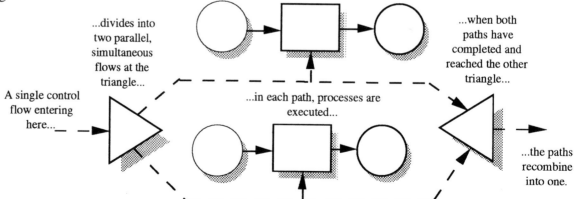

Independence of Parallel Paths

The parallel flows represented here are either physically independent — that is, are being carried on by separate processors — or represent a valid simulation of physically independent processing, such as time-slicing. What parallel flow does *not* mean, in this notation, is that all of one flow line can be processed and then all of another — that's not a valid simulation of physically independent processing. One flow line, for instance, might have a decision block which is waiting for data coming from the other flow line, and if the waiting flow line is processed first then it will hang up.

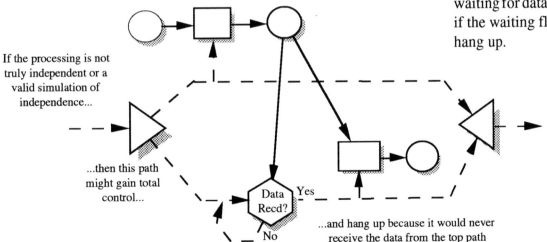

If the processing is not truly independent or a valid simulation of independence...

...then this path might gain total control...

...and hang up because it would never receive the data from the top path

Interconnected and Hierarchical Diagrams

"Flat" Connections with Page Connectors

A time-honored way of partitioning diagrams of all kinds is through the use of page connectors, and they can be used with this notation as well. They can connect both data flow lines and control flow lines that go across page boundaries. The resulting sets of diagrams do not, however, permit the user to see the "big picture," and it is easy to get lost if you have to trace through very many diagrams.

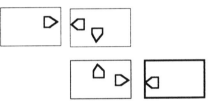

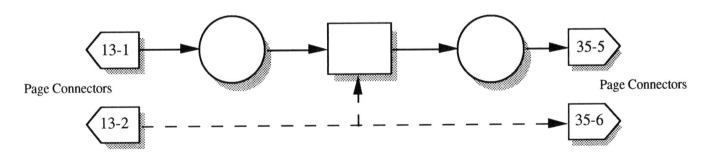

Page Connectors   Page Connectors

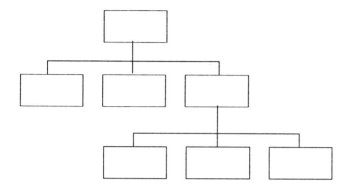

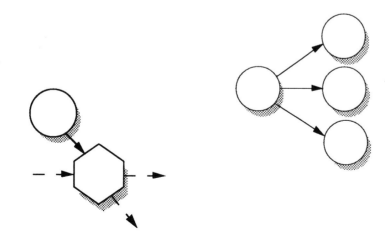

A Better Way for Many Tasks: Hierarchical Interconnection

Most modern diagramming methods for software provide a hierarchical breakdown with which entities on higher levels can be shown broken out into whole diagrams on the next level down. Typically this is done for processes, and our notation does so also. However, it provides for the same sort of breakdown for decision blocks as well. In addition, on each diagram it is possible to show data circles that really represent aggregations of elementary data items, and to show a hierarchical breakdown of those data aggregates into their elements. We'll look at each of these functions, but first let's examine why you would want to show things hierarchically.

Hierarchical Designing and Debugging

You will find, in working with this notation, that it models more of your design than other notations, and that it covers many more aspects of the work. That helps you to achieve an uncommonly deep understanding of your design.

But, because it does cover so many aspects of the work, this notation asks you to put a great deal into it; it requires real thought to do one of these diagrams properly. And in doing it properly, you are designing the product: you will be discovering all the aspects of your design that might later haunt you in testing. So it's not accurate to think of the diagramming as just a diagramming exercise, but rather as a design and debugging exercise. Then, the diagram will be an excellent tool for discussion of the software; it will provide other people the opportunity to understand your design and to comment on it. When they do that, they are participating in the design and debugging process.

In a later chapter, we present a detailed example, about a digital wristwatch, which turns out to be surprisingly complex. Suppose that you are the chief designer of the wristwatch. If you must understand what is in the wristwatch, if you must parcel the work out to individual software engineers, if you must prove to yourself that the design will work, and if you must convince others that it will work, then of course you must deal with all the complexity that is in it. If you don't deal with this complexity at the time you design it, you will deal with it later — when the watch doesn't work the way everyone wanted, and has to be fixed.

There are some other diagramming methods which don't model as many aspects of your solution, and these methods are therefore easier to deal with during the design phase. That leaves you to face the other aspects — the ones they don't model — later, when the cost to fix them is much higher.

Structuring Your Diagrams

This notation allows you to structure your diagrams in four basic ways:

- They can be structured as one large diagram, with control flow lines and data flow lines running anywhere on the diagram, as required.

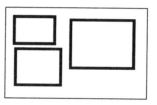

- They can be structured as one large diagram, but with enclosures or "fences" separating various sections of the diagram. Control flow lines and data flow lines still run anywhere on the diagram, crossing the enclosure boundaries as required.

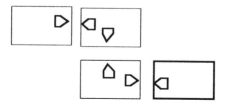

- As noted above, they can be structured as separate small diagrams, all on the same level, with control flow and data flow arrows moving from diagram to diagram via page connector symbols.

- Finally, they can be structured as a hierarchical set of diagrams, in which the process blocks are detailed out on the next level down, each as a separate diagram. In this case, control and data flow come into a lower level diagram from the diagram above it, and when they leave the lower level diagram, they go back to the diagram above it. Communication to other parts of the system is shown within an upper-level diagram at the appropriate level.

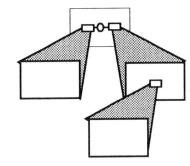

You can also use combinations of these; for instance, you can have a large diagram at the top level, and then have process blocks on that diagram detailed out as separate small diagrams.

We've elected to structure the main example in this book the fourth way, as a hierarchical set of small diagrams. This has the advantage that you can see the design developing level by level, and during development it allows the upper level diagrams to control the communication among the lower level diagrams, which helps to prevent problems.

Hierarchies of Processes

If you have a process box labelled as doing some fairly complex function, and you would like to show the processing which is going on down inside that function, we show that by providing a whole separate drawing that is cross-referenced to the block it details through the use of reference names or numbers. That drawing can, of course, include lower process blocks, data circles, control flow, decision blocks, and so on. In turn, a process box on that diagram could be detailed on a still lower level diagram, and this hierarchical structure can be carried to whatever level is needed.

In the chapter "A Detailed Example: A Digital Wristwatch," you'll find a full example of a hierarchical breakdown of process blocks.

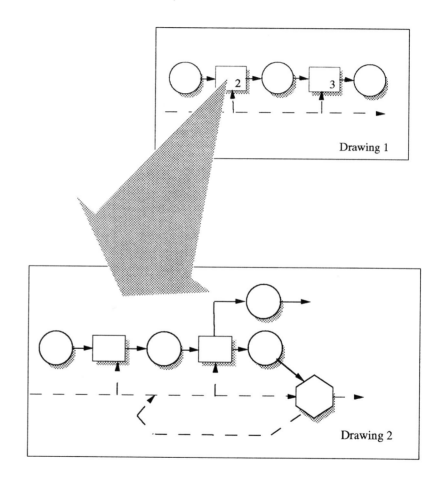

Drawing 1

Drawing 2

Data Flow Between Higher and Lower Diagrams

When a process block on a higher diagram is exploded out to a whole lower diagram, there will be data circles entering and leaving the process block on the higher diagram. These same data circles should appear on the lower diagram, with the same legends. The lower diagram should show how this data is used in the various process blocks and decision blocks found there.

On the lower diagram, if a data circle is seen coming into the diagram "from nowhere," it's assumed that it is coming from the parent diagram — the one next higher up. Similarly, a data circle going off "to nowhere" is returning to the upper diagram. Such circles may be shown either with or without the arrows that "come from or go to nowhere."

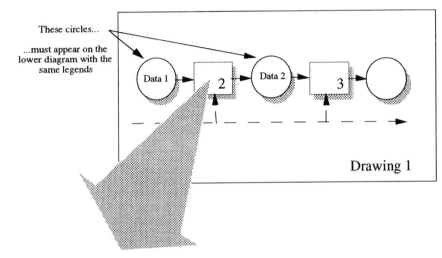

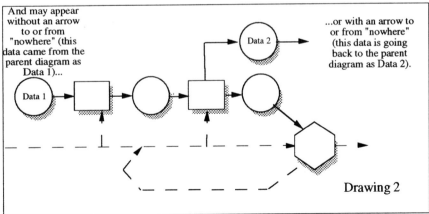

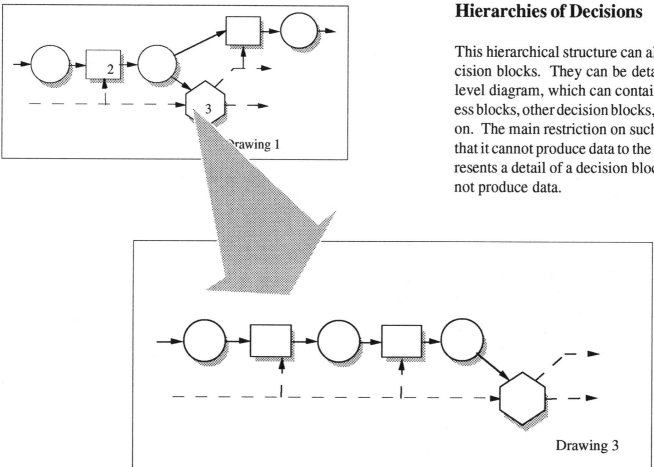

Hierarchies of Decisions

This hierarchical structure can also be applied to decision blocks. They can be detailed out on a lower level diagram, which can contain data circles, process blocks, other decision blocks, control flow, and so on. The main restriction on such a lower diagram is that it cannot produce data to the outside, since it represents a detail of a decision block, which itself cannot produce data.

In the chapter on the digital wristwatch example, section "1.2.1 Alarms?" shows a full example of hierarchical breakdown of a decision block.

Hierarchies of Data

At times it will be convenient to show a circle as containing an aggregation of data, especially when data enters or leaves a diagram. Then it will be useful to be able to show that aggregation of data being broken out into its components, so that each can enter into various processing blocks. In this notation, the aggregated data is shown as a circle, and then solid data flow lines are shown extending from that circle to all the component circles.

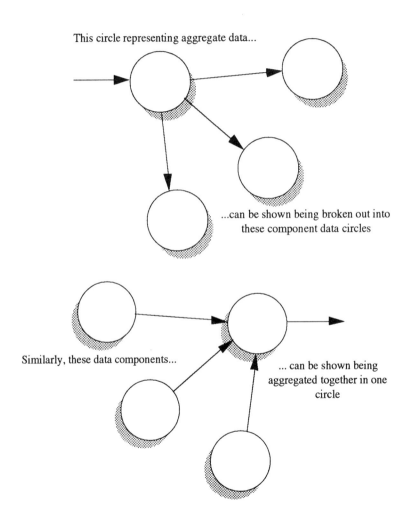

This circle representing aggregate data...

...can be shown being broken out into these component data circles

Similarly, these data components...

... can be shown being aggregated together in one circle

A Detailed Example:
A Digital Wristwatch

As a way of explaining the notation further, we used the notation to try to "reverse-engineer" the mode functions of a digital wristwatch, just by observing the way the watch responds to its control buttons. The resulting design may not represent the way the watch was actually designed, but it should represent a design that has the same external functional attributes as the watch we used:

• Normal time, including year, month, date, day of the week, hours, minutes, and seconds;

• "Dual" time, that is, hours, minutes, and seconds for a second time zone;

• A settable alarm that will go off at a particular month, date, hour, and minute;

• A settable count-down timer, with an alarm at the end;

Mode	Display
Normal Time	12:30:36
Alarm	7:15
Dual Time	9:30:36
Timer	00:15:38
Chronometer	03:41:89

- A chronometer or stopwatch, which is essentially a count-up timer.

The wristwatch has one "mode" for each of the above functions, plus some sub-modes under each of those, and is operated by four buttons. Button #4 is usually used for changing from one mode to another; button #1 is usually used for sub-modes where times are set; and button #2 is usually used for turning on a light on the watch. But the meanings of all the buttons can change depending on what mode the watch is in, especially button #3, which has different functions in most of the modes. In fact, the various states and modes are quite complex, and when they are described carefully and accurately with our graphical notation, we can see that they aren't at all trivial. The diagrams are quite involved — but that's because the watch itself is involved, and that is the way the real world is. You should ask yourself: can I really tell how states and modes of the watch work from these diagrams?

1.0 Wristwatch

Let's look at a top level diagram of the watch, which we've designated "1.0 Wristwatch." We guessed that this particular watch, like many small electronic systems, probably has a number of parallel, independent functions that communicate with one another. Here we show the flow of control, which enters where the word "in" is noted, dividing into four parallel paths, each of which then loops indefinitely. One path does the display and control functions, another rings the alarms, another does the basic time measurement, and the last controls the light. Because they loop indefinitely, each at its own rate, the paths are not shown coming back together in a closing triangle, nor is there an "out" arrow showing control leaving this diagram.

It's true that parallel functions, operating at different rates, and sharing information, can get into trouble if one function tries to read information at the exact instant that one of the others is trying to change it. That requires complex signalling to prevent trouble; we'll describe how that can be handled later in the book, but for this example we have neglected that effect.

Of course, the watch wouldn't have had to be designed in a parallel manner. For instance, all the functions could have been done sequentially rather than in parallel. We didn't think they had been, because the chronometer time has to be updated once every 1/100 of a second, and can't be neglected while the watch attends to (say) the changing of the timer time setting.

But the manufacturer could also have chosen an interrupt-driven design, and later on we'll show what this top level chart might have looked like if that were the case.

From this top diagram you can tell some things at once: for instance, the buttons allow you access to the display and control functions and the light control, but not to the time measurement or the alarm ringing function — except through messages passed from the display and control. So that's a significant aspect of the design of the watch as we've guessed it to be. The diagram shows this distinction clearly.

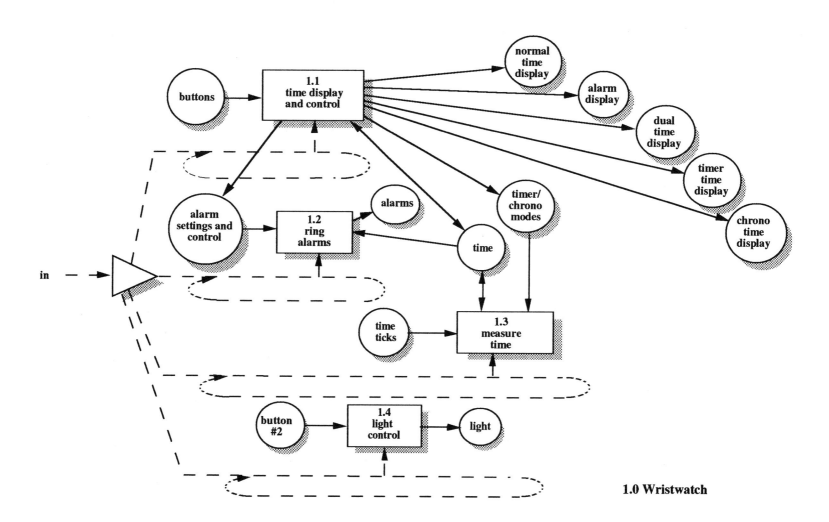

1.0 Wristwatch

A Detailed Example: A Digital Wristwatch

In this set of diagrams of the parallel solution, as mentioned above, we've numbered the top diagram "1.0." Each of the process boxes on this diagram will be detailed out on another diagram, and we've given them the numbers 1.1, 1.2, and so on, so you can find those diagrams.

Going down the chain, let's look at 1.1, Time Display and Control.

1.1 Time Display and Control

One of the rules of the notation is that, if a data circle enters a process block on an upper-level diagram, that same data circle (with the same legend inside it) must also be shown on the lower-level diagram that shows what is inside the process block on the upper-level diagram. This is sometimes called "balancing" the diagrams, and the checking of this is a primary feature of most diagram-based CASE (Computer-Aided Software Engineering) systems. We've been careful to follow that rule here:

At the top of diagram 1.1, we've shown the data circle "time"; on the upper-level diagram 1.0, we saw that this circle came from the time measurement function. Now, on diagram 1.1, we see that "time" is a composite data item that has four components: normal time, dual time, timer time, and chrono time. We've shown this by drawing solid arrows from the "time" circle to each of its component circles. Each of these component times, of course, is based on minutes and seconds, but they have different starting values and maximum ranges and so are separate data items. They will be displayed on the watch by various functions that appear on diagram 1.1. But the "time" circle is also shown as leaving the diagram; that's because the various components can be reset or changed by functions in diagram 1.1, and the changed values will return to the time measuring function.

1.1 is therefore the diagram that shows the five main modes of the watch. These modes are shown as being controlled in a loop which is unending once it is entered, which is the way the watch seems to work as long as the battery is functioning. In three of the modes, you can go to the next mode in sequence or

A Detailed Example: A Digital Wristwatch

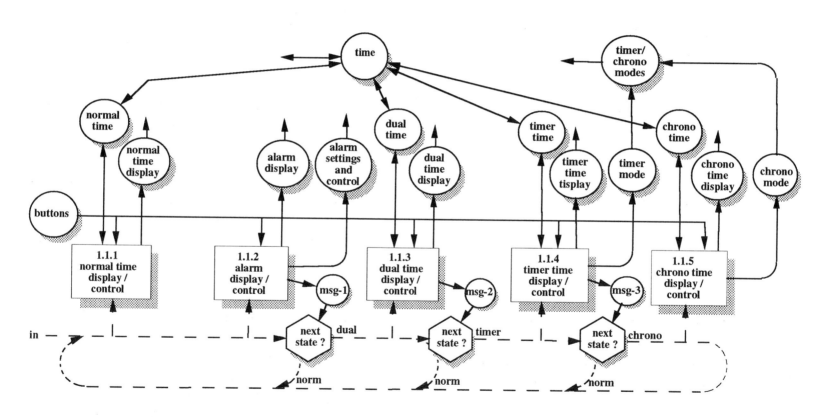

1.1 Time Display and Control

A Detailed Example: A Digital Wristwatch

back to the normal time display. This is controlled from inside the modes by a message (for instance, "msg-1" on the diagram) that comes out of each one where it's needed.

If the watch manufacturer had designed all of the functions of all the modes on one big diagram, moving from mode to mode could be done directly without using these messages. Sometimes that is desirable, and we'll show some examples of that later in the book. But if the manufacturer wanted each mode to be worked on by a different programmer, then showing the various modes separately like this is better practice. Communicating between the 1.1 diagram and the various individual modes by messages is then the right way to do it, according to most books on modern programming practices. So we've shown it that way.

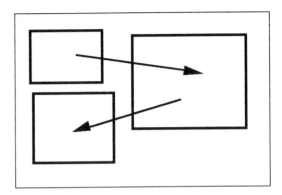

On 1.1, button signals enter each of the modes, and displays are produced from each one. Note that each of the modes produces information that ultimately will go to the other major watch functions (1.2, Ring Alarms, and 1.3, Measure Time). In the hierarchical

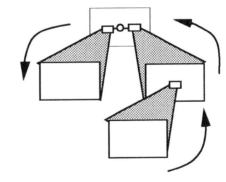

form of the notation, the circles representing that information are shown with arrows just ending in midair; this means that the information is being passed back up to diagram 1.0 (its parent diagram). It is on that upper level diagram that you can trace where the information goes (such as 1.2, Ring Alarms). For this to work, information circles leaving the lower level diagram (1.1, in this case) must appear on the upper level diagram (here 1.0), which is why that's required by the notation.

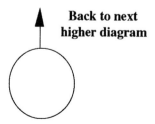

1.1.1 Normal Time Display / Control

"Normal" time is what the watch ordinarily will be displaying, as distinguished from "dual" time (which is usually time in another time zone). It displays normal time in either the ordinary civilian format or the 24-hour military format. The watch also displays a "reminder," which is just a mark that the wearer can use to remember something, and which the wearer can set and reset.

Here we've broken the normal time display function into three process blocks, including the subfunctions for setting the normal time and for turning the reminder on and off. The main function "display normal time" could, of course, be detailed below this point, but we have elected not to for this example. Notice the data circle "normal time 24-hr mode" which passes between 1.1.1.1, the Normal Time Set function, and the "display normal time" function.

Note that when button #4 is detected, control leaves this diagram, which means that it returns to diagram 1.1. On diagram 1.1, control is then shown moving directly to diagram 1.1.2, Alarm Display/Control.

Button #1, however, causes control to enter the process block 1.1.1.1, Normal Time Set, in which the setting of Normal Time can be changed.

50 ───────────────────────────────────── *A Detailed Example: A Digital Wristwatch*

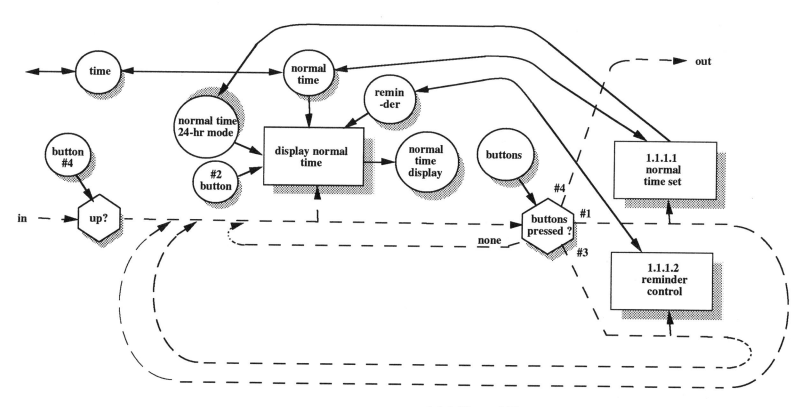

1.1.1 Normal Time Display/Control

1.1.1.1 Normal Time Set

The first decision block in this diagram simply tests to see whether button #1, which was used to enter the normal time set mode, has been released. Although the display associated with Normal Time Set appears as soon as button #1 is pressed, other functions in Normal Time Set don't become available until button #1 is released. To show this, we invoke a special use of the decision block: when only one dotted arrow emerges from it, the action is that control waits inside the decision block for the correct data to appear, which in this case is the absence of a button #1 signal. Control then advances beyond the decision block. This kind of button interlock mechanism is used in three other places on the diagram.

Here on 1.1.1.1 we have used the triangles to indicate parallel action: the Normal Time setting is continually displayed while other actions on the diagram are carried out. We've guessed that the display is done on one parallel path, which enters a loop that repeats the display. The other parallel path enters a decision block leading to the other functions. Button #1 releases the display loop, and simultaneously provides an escape from the loop doing the other functions; the two paths come together in the closing triangle, and the resulting single path then exits the diagram, back to the parent diagram 1.1.1.

One subfunction on this diagram is used to select an "aspect" of normal time, that is, the year, the month, the date, the hour, or the minute. Each depression of button #4 selects the next aspect; after minute is selected, the next selection will be the year again. We could draw a diagram on the next level down which detailed that action, but have chosen not to do so. However, once an aspect has been selected, then button #3 causes that aspect to be increased repeatedly as long as button #3 is held down. We see that through the loop path back to "increment selected aspect." When any aspect reaches its maximum (such as minutes reaching 60), it automatically resets to its minimum; if we wanted to show those details, they would be shown on a diagram which details the process block "increment selected aspect."

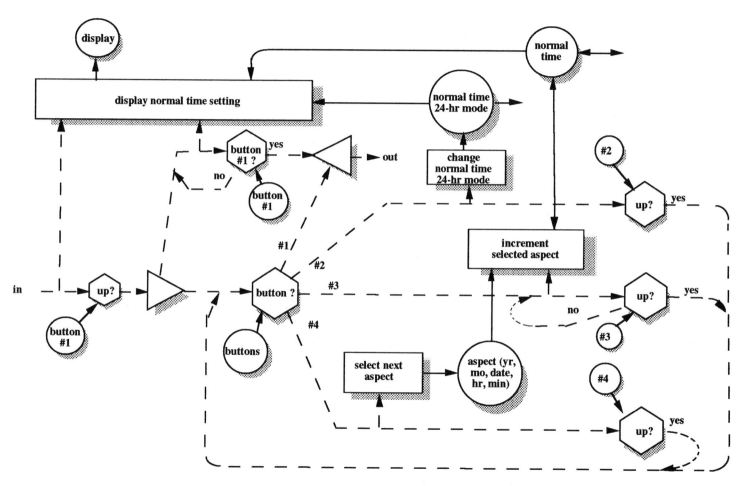

1.1.1.1 Normal Time Set

1.1.1.2 Reminder Control

We've included the "Reminder Control" diagram for completeness, although it is the simplest diagram of the set. Of course, that's a reflection of the simplicity of the reminder function itself.

1.1.2 Alarm Display/Control

This diagram shows the same parallel path structure that we saw in 1.1.1.1, Normal Time Set. Three functions of the watch are present here. The first is the setting of the time the alarm will sound. The second is the setting of the chime (the alarm sound itself) so it will sound either on the hour, or when the alarm goes off, or both, or neither. The third is just a test of the chime sound, presumably as a check on battery power remaining.

And, here in this diagram, we have the first case in which we must control whether the watch will return to the normal time display or go on to the next mode,

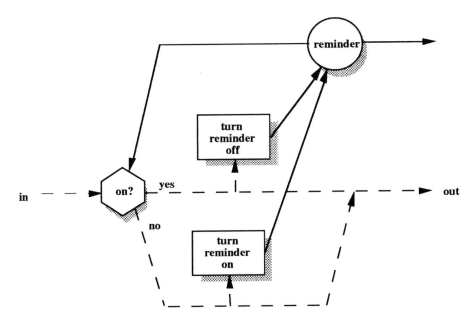

1.1.1.2 Reminder Control

54   *A Detailed Example: A Digital Wristwatch*

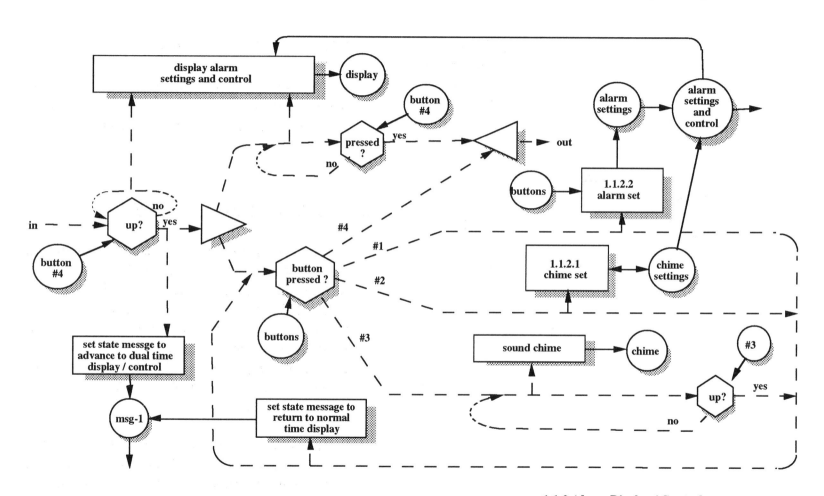

1.1.2 Alarm Display / Control

which in this case is the dual time display. In our design, we have elected to put the decision blocks which control that on diagram 1.1, Time Display and Control. But we control those decision blocks by messages sent from within the individual modes, such as 1.1.2, Alarm Display/Control.

On this point, it is worth stating again the options that you have if you are the chief designer of the watch. If you want to show the whole watch on one big diagram, then these messages are not necessary. You can just use different decision blocks to direct control flow to the dual time mode or back to the normal time mode. The diagramming job is then easier. However, then the picture that you now see condensed onto diagram 1.1 is spread out over the whole big diagram. You also don't see the data interfaces condensed onto one high-level diagram. As the chief designer of the watch, you may then have a harder time seeing whether they are correct.

With smaller diagrams arranged in a hierarchy, you have control of major control flows and major data flows at a high level, you can partition functions out to individual programmers more easily, and each individual programmer has a cleaner set of interfaces with which to work.

So, here we have shown the hierarchical approach, which requires that messages be produced by the lower level processes (such as 1.1.2, Alarm Display/Control) and fed up to the higher diagram (1.1) where the control flow is managed. Here, message "msg-1" is generated from within 1.1.2. If button #4 is the first button pressed after button #4 is released, then the message will say that the watch should go on to the dual time mode; but if some other button is the first pressed, the message will say that the watch should go back to normal time display.

1.1.2.1 Chime Set

This is a simple function which is driven from button #2 in 1.1.2, Alarm Display/Control. It advances the alarm/chime control through four states: alarm chime, hour chime, both, or neither.

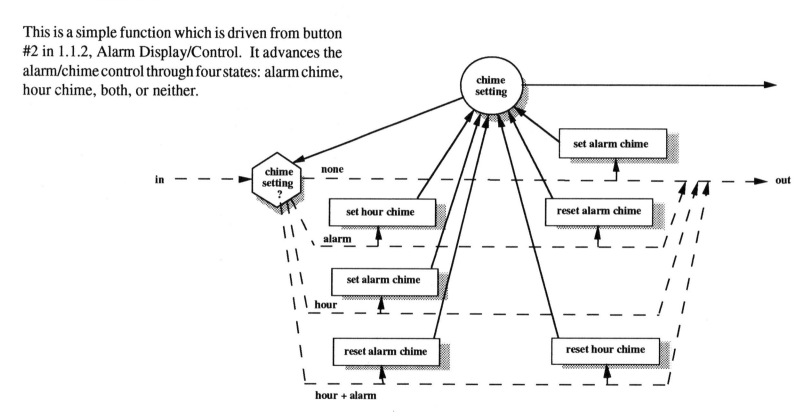

1.1.2.1 Chime Set

1.1.2.2 Alarm Set

The Alarm Set function works almost the same way that the Normal Time Set does, by allowing the selection of an aspect (month, day, hour, minute) and then allowing that aspect to be incremented. Since it's driven from one of the parallel paths on 1.1.2, Alarm Display/Control, it operates in parallel with the display of the alarm setting, so that the user can see what is happening while the numbers are being changed.

1.1.3 Dual Time Display/Control

This mode displays dual time, which is usually time for another time zone. However, both hours and minutes can be set, and on this watch you can set the 24-hour military time option separately from the one which affects normal time, so that you can have normal time in both civilian and military formats if you like. This diagram has the parallel display loop, the button #4 interlock, and the state message feature that we saw in Alarm Display/Control.

1.1.3.1 Dual Time Set

This diagram is almost like that of 1.1.1.1, Normal Time Set, except that here we have not shown the parallel path for the display, since 1.1.3.1 is driven from one of the parallel paths on 1.1.3, Dual Time Display/Control, on which the other parallel path takes care of the display. However, here only the hour and the minute can be changed, not the year and month.

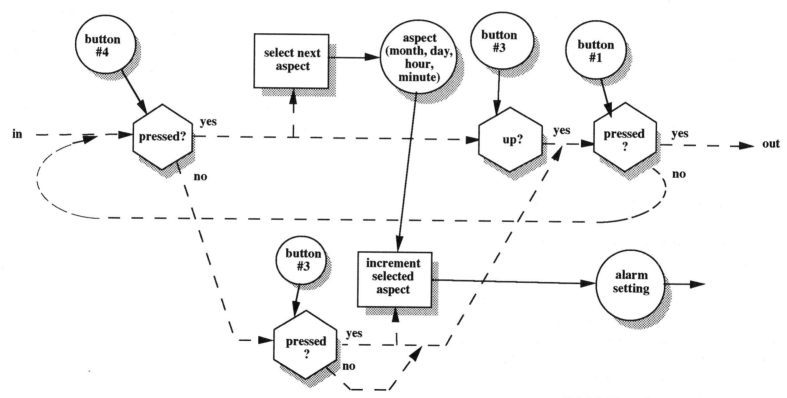

1.1.2.2 Alarm Set

A Detailed Example: A Digital Wristwatch

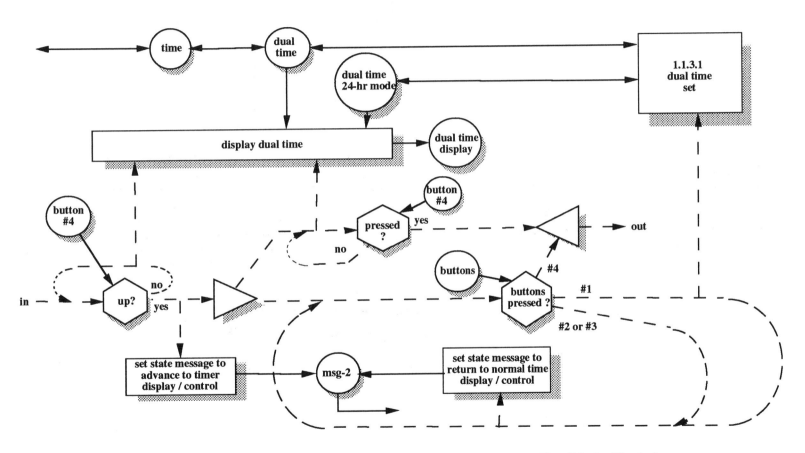

1.1.3 Dual Time Display/Control

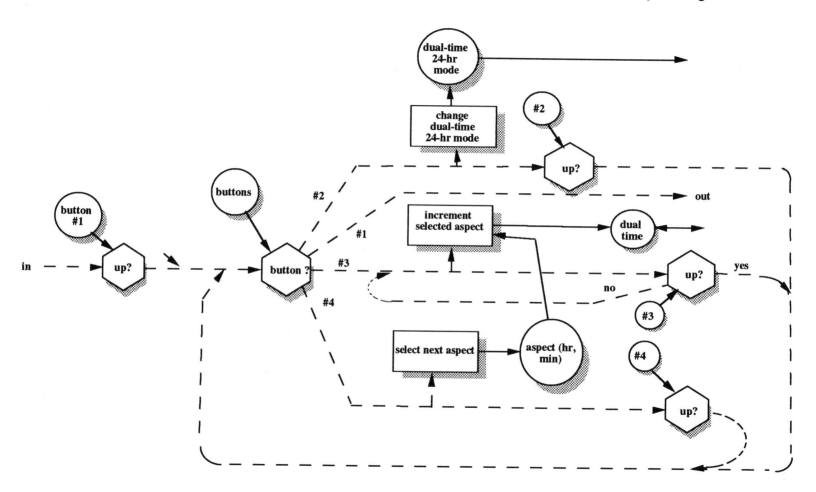

1.1.3.1 Dual Time Set

1.1.4 Timer Time Display/Control

The timer is able to count down from any setting up to 23 hours, 59 minutes, and 59 seconds, and will ring the alarm when zero is reached. This diagram is mostly composed of features we have seen on the other diagrams, except for the toggling on and off of the "timer mode" by button #3. The timer mode is a message to 1.3, Measure Time, to start or stop counting down the timer time. This message is passed back up to 1.1, Time Display and Control, where it is combined with a similar message from the Chrono mode, and sent up to 1.0, Wristwatch. On that diagram, you can see it being sent to 1.3, Measure Time.

1.1.4.1 Set Timer Time Setting

Again, this diagram is almost the same as the other diagrams which select and increment aspects of one of the time components.

1.1.5 Chrono Time Display/Control

"Chrono" means chronometer, or stopwatch; this mode is essentially a count-up timer. Button #3 allows the user to stop and restart the timing, thus giving "lap times"; this is handled through the chrono mode message, which instructs 1.3, Measure Time, to start or stop the incrementing of the displayed value. Button #2 resets the chrono time to zero. The diagram does have the button #4 interlock.

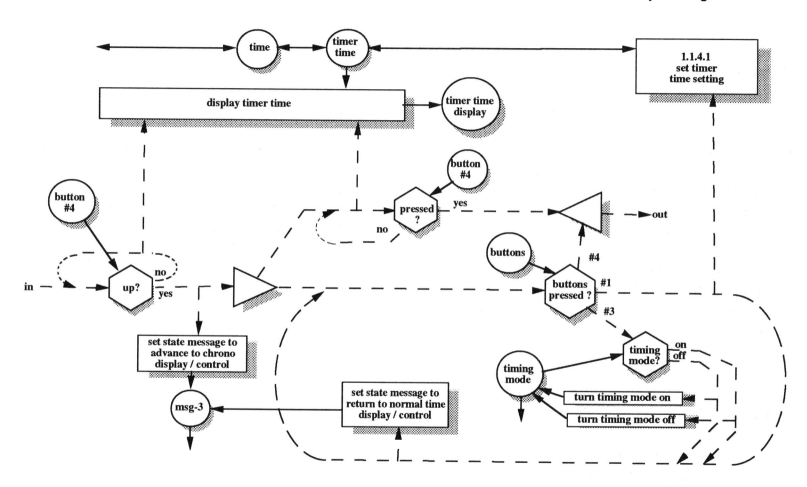

1.1.4 Timer Time Display/Control

A Detailed Example: A Digital Wristwatch

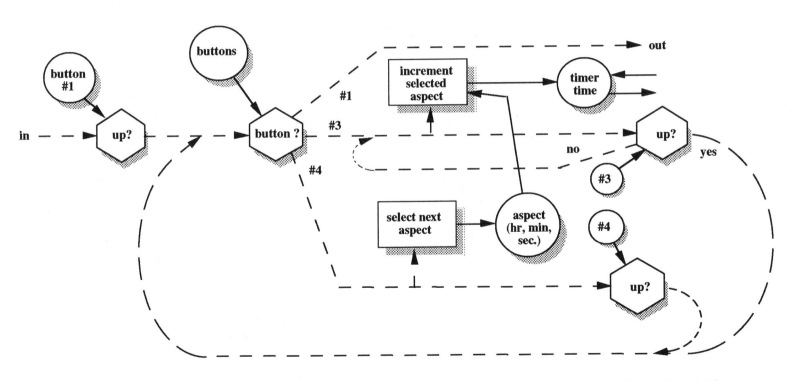

1.1.4.1 Set Timer Time Setting

A Detailed Example: A Digital Wristwatch

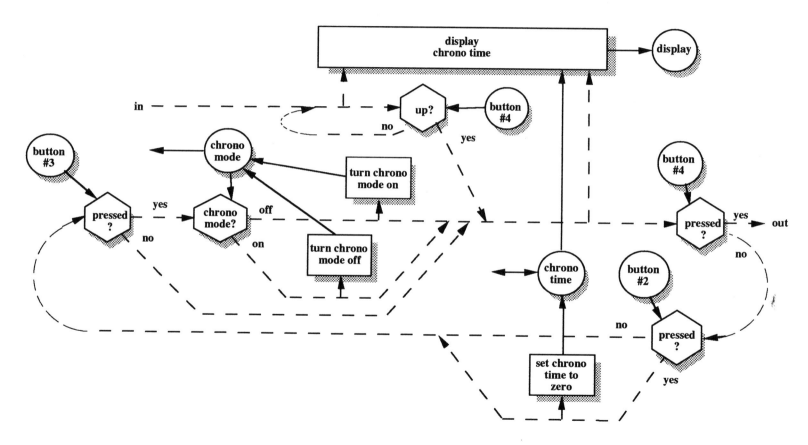

1.1.5 Chrono Time Display / Control

1.2 Ring Alarms

On diagram 1.0, Ring Alarms is shown controlled by a separate parallel loop, and fed by the data "alarm settings and control" and by "time." The Ring Alarms function begins by testing to see whether there are any alarms to be rung, and if so, sounding the alarm chime. It continues to sound the alarm until there is a button actuation, or the end of a timeout period occurs. Then it goes on to test whether the hour chime needs to be rung.

Here in this diagram, we have a decision block which is detailed out on a separate diagram (1.2.1, Alarms?). The rules of the notation specify that a diagram detailing a decision block cannot produce data, and it must produce the same number of outgoing control flow arrows as the decision block it details. We'll now see how this works:

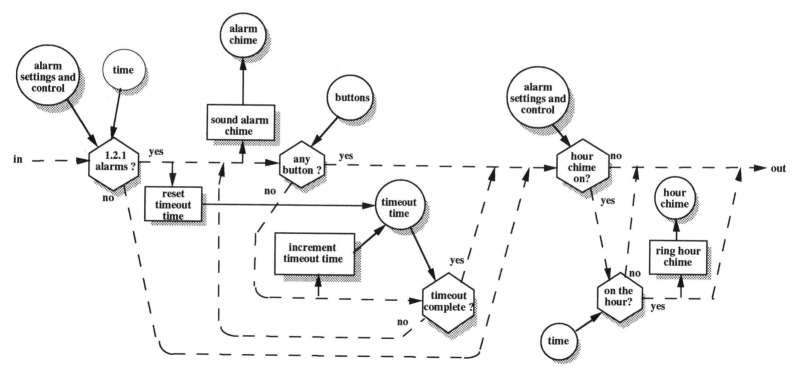

1.2 Ring Alarms

A Detailed Example: A Digital Wristwatch

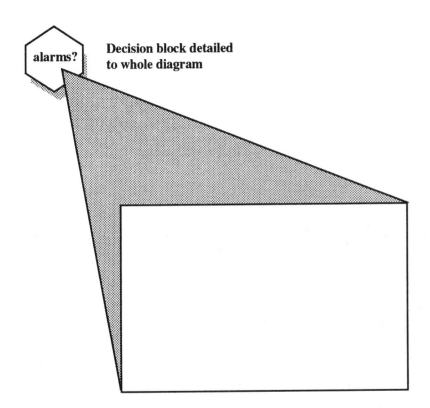

Decision block detailed to whole diagram

1.2.1 Alarms?

Note that this diagram consumes data but it does not produce any, and it produces two control flow outputs, one each for the "yes" and "no" output of the decision block it details. This diagram also illustrates the splitting of aggregate data circles into their component data items.

68 *A Detailed Example: A Digital Wristwatch*

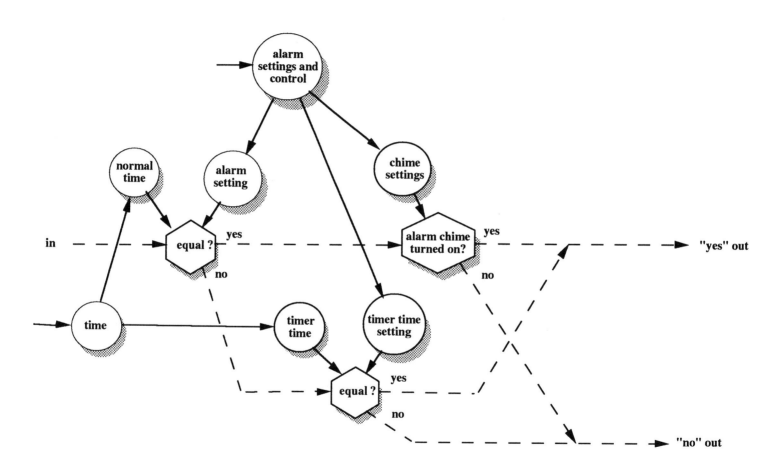

1.2.1 Alarms?

1.3 Measure Time

This is the diagram that actually measures time, given an input time tick, then provides that time to the display functions we have been considering. We have detailed out the minute, hour, day, month, and year checks for normal time by one more level, in 1.3.1. The checks for dual time are different, since dual time does not have day, month, and year.

A Detailed Example: A Digital Wristwatch

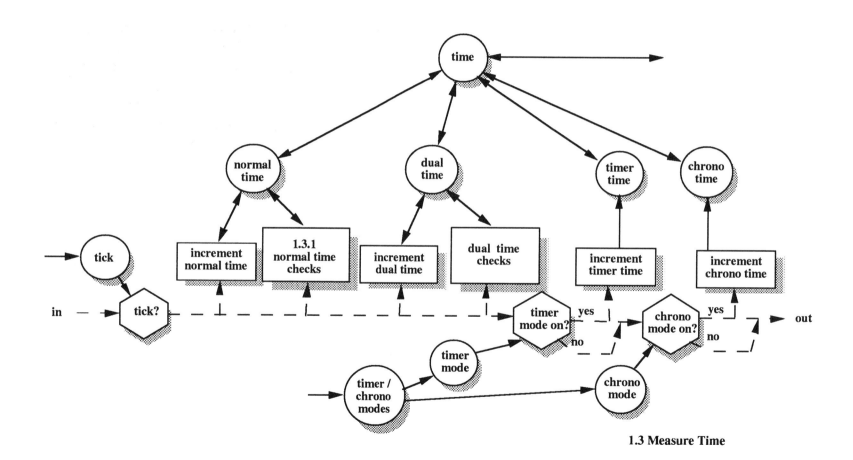

1.3 Measure Time

1.3.1 Normal Time Checks

From this diagram, you can tell that we have guessed that the checks and adjustments for end of minute, end of hour, end of day, and end of year are done computationally, while those for end of month and for leap year day are done with tables. We also guessed that the day of the week is calculated rather than looked up in a table. These functions could be detailed out one or two more levels, or it might be decided that a transition to code would be made at this level. That decision would be based on how readable the code could be made, and whether the code itself would be adequate for discussion and design reviews at this level.

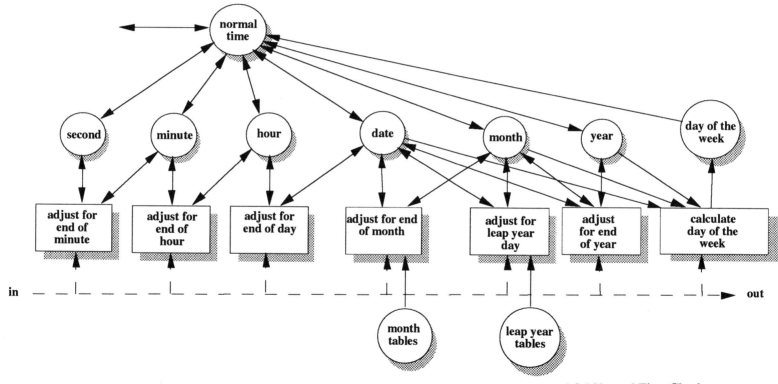

1.3.1 Normal Time Checks

1.4 Light Control

This diagram is included for completeness, although it is extremely simple.

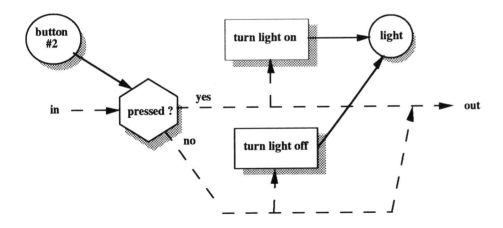

1.4 Light Control

Separation Through Enclosures

Enclosures

Diagrams composed of the icons and interconnecting lines described above can be surrounded or partly surrounded by enclosures for a variety of purposes. These add meaning without subtracting from the ability of the notation to show processes, data, control flow, and decisions all at once. Through the use of enclosures, the diagrams can be arranged to show the partitioning of software into states and state transitions, software objects, physical CPU's, and other useful partitionings. In general these enclosures, even when used for different purposes, can be nested inside one another in any order that is required. We'll describe below a number of these enclosure types.

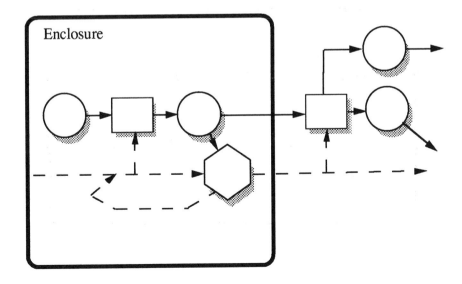

States and State Transitions

When an enclosure is used to indicate a software state, the diagram is partitioned so that dotted control flow arrows carry the control from one state to the next. Solid arrows carry data from one to the next; the data circles can be shown either inside a state or in the area between states. Instead of the "state 1" and "state 2" notations shown here, you can use names identifying what the states do.

Decision Blocks and State Transitions

Decision blocks within states show the conditions which lead from one state to another, that is, which lead to state transitions. Thus all the information that exists on a state transition diagram is available, along with the other information available in this notation.

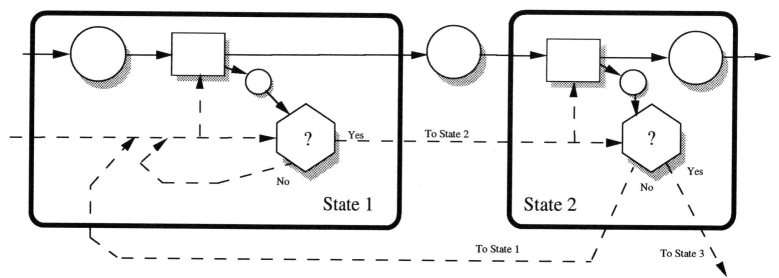

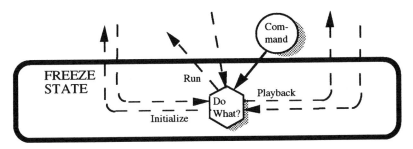

A Passive State

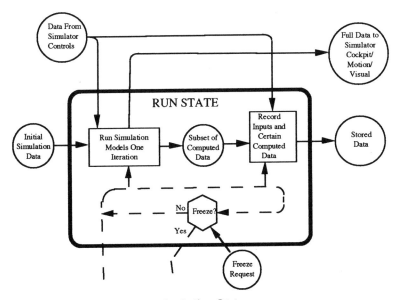

An Active State

Active and Passive States

If nothing but a decision shows inside the state, then it is a passive state which does no actual processing other than accepting inputs and switching to another state. However, states can be active, having process boxes within them which carry out work. Active states can, of course, contain decision blocks which direct control flow to other states.

Comparing these diagrams with a state transition network, we see that the individual states show in both types of diagrams, as do the conditions under which transitions take place. Sometimes actions taken are shown on state transition diagrams. Here, all of those aspects are shown, as well as the data which flows between states and within states.

A Flight Training Simulator

The state transition diagram shown on the next page is taken from a flight training simulator, and includes an active "run" state, in which the student pilot is flying the simulated aircraft, and the inputs (and certain computed variables) are being recorded as data points. On request, one can make a transition to a passive "freeze" state, and from there to any of three other states: an initialization state, a playback state in which the student pilot can see his mistakes replayed, or back to the run state.

These states are all active except the freeze state, which contains only a decision block that directs control flow to one of the other states when an appropriate command is received.

States in a flight training simulator are typically more complex than shown here.

Each of the process blocks shown in these states can be exploded on its own diagram exactly as in the other examples in this book.

Separation Through Enclosures

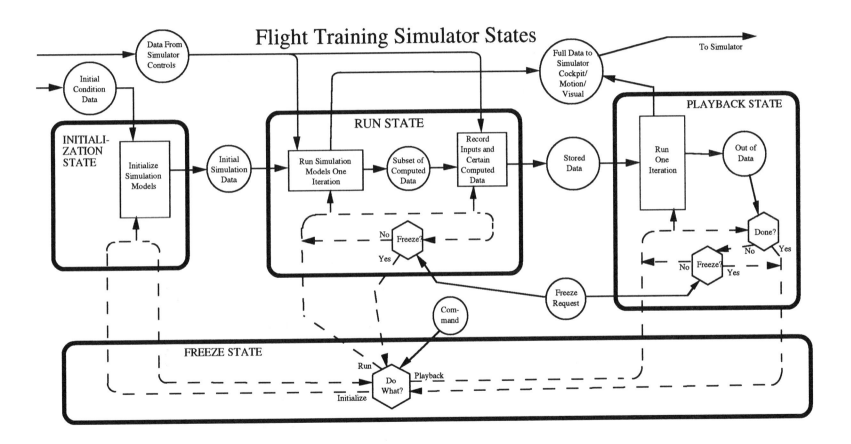

Physical CPU Enclosures

A simple enclosure can also be used just to show that a particular piece of data, executable code, or both exist within some particular processor, and to show data flowing from one processor to another. This usage would be similar to the use of enclosures on electrical diagrams to show the various boards and cabinets containing the circuits.

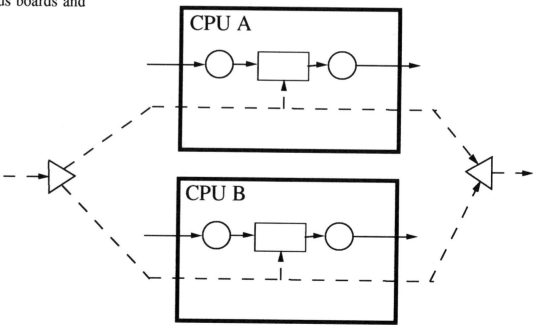

Software Objects

Object Enclosures

Enclosures can also be used to show software "objects." Objects consist of both data and executable code. Some of the data may be accessible to software outside the object, but usually an object will have some internal data which is to be protected from outside tampering. Then there will be a set of "operations" which can be accessed from the outside and each of which carries out some work on the protected data — such as obtaining a copy of part of it for use by some outside code, effecting a change to some of the data, and so on. The data is protected because only these operations are allowed to act on it. In some books on the subject, the operations are called "services" of the object. Textbooks on object-oriented analysis and design often refer to this data protection as "encapsulation," which is one of four properties identified with software objects.

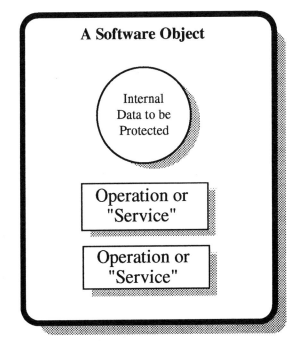

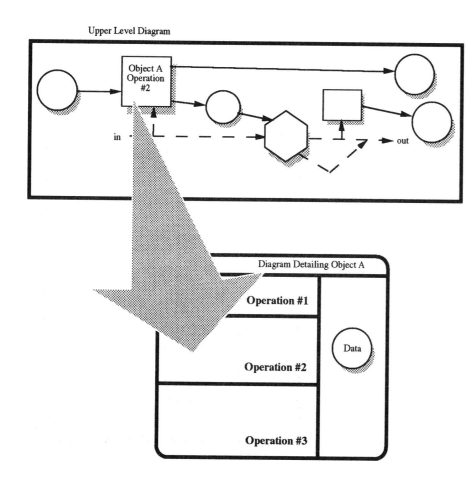

Operations of Objects

Operations of objects are processes. An operation can be represented as a process block an any diagram (including a diagram of another object) which makes use of, or "calls," the operation. In showing the explosion or hierarchical detailing of such an operation of a software object, however, we can show the detailing of that process block *together with* the detailing of the other operations of that object, in a single diagram if they will fit. We can enclose them all together, and identify them as the object; but in doing so, we also separate the operations from each other (with lines) and from the data which is protected by the object. In that way, it becomes clear where the boundaries of each operation lie, but it also becomes clear that all of these operations are associated with the one object. The way in which all operations interact with the protected data will be displayed. Note that an operation of an object can be called from several places, in each of which you would find a process block labelled with the name of the operation.

Software Objects

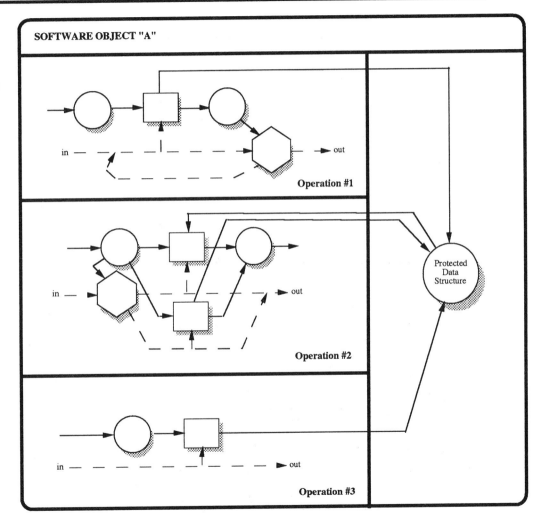

—Any drawing, even of another object, can make use of, or "call," an operation of an object, by showing it as a process block labelled with the operation's name — here "Operation #1."

Here, ⟶ within the object, is this detailed drawing, or exploded view, for Operation #1 ...

... which is "packaged" next to the detailed drawing for Operation #2 ...

... and Operation #3. All three operations read or modify the same data, which is shown in the right compartment of the software object.

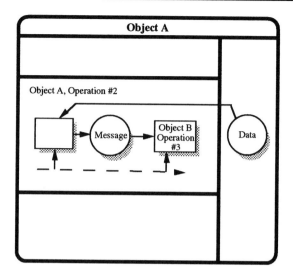

Here, a message is generated and then passed to Operation #3 of Object B. That operation is shown here as a process block labelled with the name of the operation. Showing it here like this is the "call" to that operation of the other object.

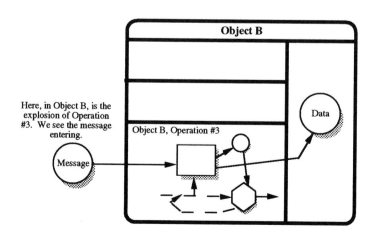

Here, in Object B, is the explosion of Operation #3. We see the message entering.

Objects often "pass messages" to one another; this is intended to isolate the objects from one another more than, say, the sharing of common data would. Still, all data is shared in some sense; it is placed in a memory location by one process, and read from that location by another. In our notation, we see those messages primarily as data circles. We show the "call" to the operation of the second object as a process block within the first object, containing the name of the operation of the second object.

A "purist" in object-oriented methods might argue that in message passing you would not invoke the other operation directly, but would send the name of the desired operation in a data circle. Then, the purist would say, there must be a special process within the second object which reads the message and uses that information to call the correct operation. This added level of indirection does help to isolate objects from one another and help to prevent unintended contamination of an object, but, of course, at the expense of added execution time. If you wish to treat the messages in this way, you can show the special operation explicitly in your diagrams.

Software Objects

In textbooks on object-oriented analysis and design, message passing is regarded as one of the four main properties of objects.

In object-oriented design, objects "inherit" operations from other objects; "inheritance" is another of the four properties of objects. This just means that when some object is asked to perform an inherited operation, it knows enough to refer the request to a second object, where the operation actually resides. Typically this second object represents a class to which the first object belongs, and the operation which the first object is invoking is some function that all the members of the class would need.

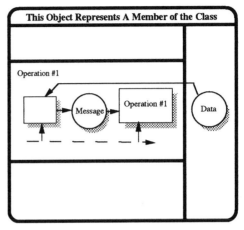

This is the object representing a specific member of a class (like a car is member of the class of vehicles). If the object needs to perform Operation #1 (an operation that is common to the whole class), you can put the operation in the object for the class, and just call it from here

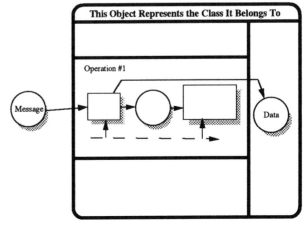

Here, in the object for the class, is where the operation is really executed. Now it needs to be stored in only one place for all members of the class.

Actually, however, there are several kinds of inheritance: one kind occurs at the time the programmer designs the object, and the programmer just copies some existing operation into the object. We could call this "design-time" inheritance, though of course some people would say that's not inheritance at all.

Then there is compile-time inheritance, in which the copying is done during compilation. In some languages, like Ada, there is a step called elaboration between compilation and execution, and inheritance can also occur there.

Finally there is run-time inheritance, in which a request for an operation really is referred to another object during execution. This progression from design-time inheritance to run-time inheritance is sometimes called a progression from "early binding" to "late binding," and late binding is also considered one of the four hallmarks of objects.

Software Objects in Industrial Software: Control of Software Changes

In the example which follows, we will stress the encapsulation theme. In military flight training simulator work, this theme is paramount because of the need to keep the simulator "concurrent" with the aircraft — that is, to have the simulator always reflect all the changes that have been put into the real-world aircraft, so that training can be as relevant as possible. To do that, it's important to be sure that each change originating in the design of the aircraft will affect one, and only one, place in the design of the simulator. Encapsulation helps with that: the simulator software is designed as a group of objects, each corresponding to some part of the aircraft. Each object is as independent as possible. Then, when a change is needed, it can be localized to the one correct place in the simulator software.

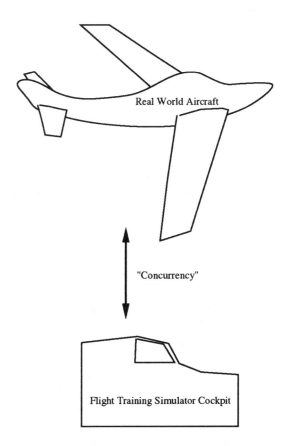

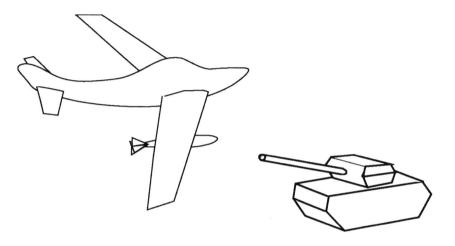

Software Objects in Flight Training Simulation

In a military flight training simulator, various real-world objects are simulated: for instance, a fighter aircraft, a missile which it fires, and a target. Software in the simulator computers represents each of these entitities. The student pilot sits in a cockpit, the instruments and controls of which drive, and are driven by, that software. In his windscreen he sees a visual scene which is computed to show exactly what he would see under real operating conditions.

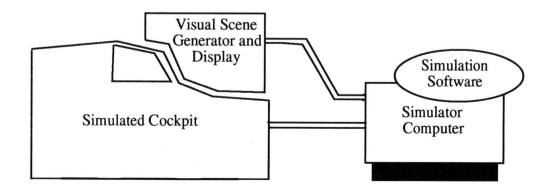

Software Objects

As noted above, one of the goals of simulation software is that when one of the simulated real world objects changes, the changes to the simulation software should be localized together. To achieve that, we shouldn't place the software that simulates that real-world object in several places; we should gather it together in a software object that does everything that has to do with that real-world object.

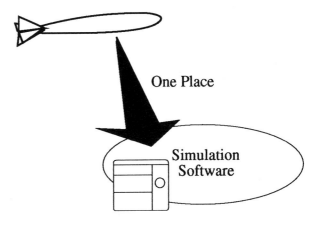

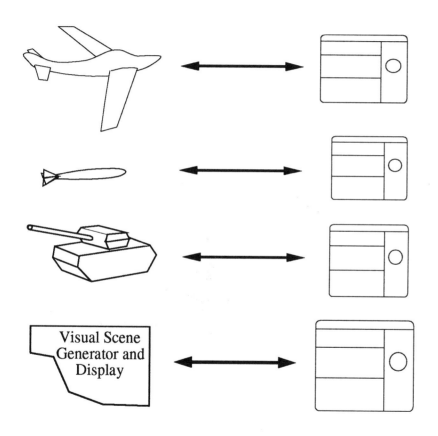

The decisions as to what real-world objects will be represented by software objects in the simulation are quite complicated and involve many tradeoffs. We'll describe one selection, but many others are possible. Here, we provide a software object for the fighter aircraft, another one for the missile, and a third one for the target. We also provide software objects for parts of the flight simulator that have no counterpart in the real world, for instance, the visual system that provides a visual scene to be displayed in the cockpit of the simulated aircraft.

Of course, the fighter aircraft has many parts — airframe, engines, instruments, controls, avionics, and so on — and we will want to have separate objects for each of those, because they can change individually. But we can have one main object that represents the aircraft as a whole.

Software Objects

Let's look at the relationship among the fighter aircraft object, the missile object, and the target object. Again, the way in which we arrange these relationships is a subject for tradeoffs, but here is one possible way. First, the aircraft tracks the target; during this phase, the target software object provides its location to the aircraft software object, and certain other characteristics so that the aircraft object can display the target's "blip" on its radar screen.

Next, the fighter identifies the target to the missile, and launches the missile. Now the missile object must request the target's location from the target object.

Finally, the missile decides that it is near enough to the target, and explodes itself. First, it can send a message to the simulator visual system that an explosion should be shown at its location. It can then send a message to the target object that it exploded at such-and-such a location, and the target object can decide whether it has been destroyed. If it decides "yes," it can then send a message to the visual system containing new instructions for the way it now looks.

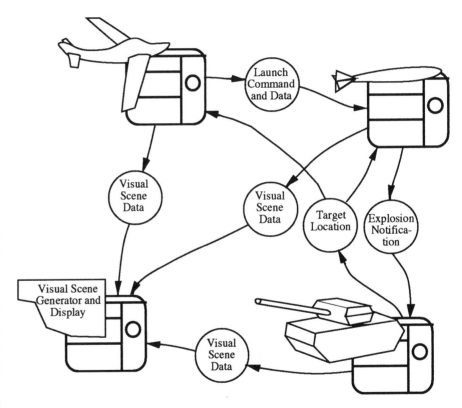

Alternatively, if the missile has a radar in it, there may be a missile radar software object, which produces a simulated beam; it may send information on that beam out to various potential targets, with a request that the target objects send messages back when they are detected by the beam. Then the radar can inform the missile when it is close enough to the target.

Clearly, we can go on adding objects for smaller and smaller pieces of the problem, and there is a point beyond which it is not sensible; for instance, when the missile explodes, we don't add objects for each of the fragments. We trade off the good that the extra objects will do in localizing the effects of changes, against the rise in overhead that the object incurs.

Software Objects

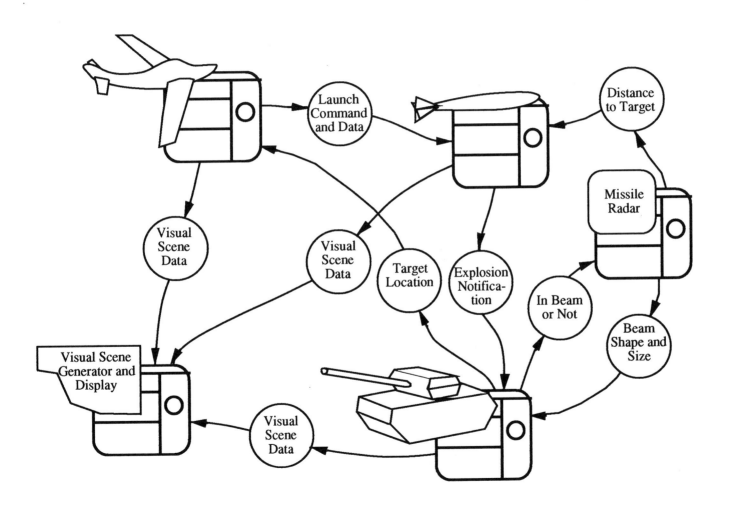

Let's look in detail at the missile object. If the objects and their relationships are structured the way we have described, we might have a process block within the aircraft object called "launch missile"; the explosion of that process block would be packaged with the missile object. It deals with data there like the location of the missile, which initially is that of the aircraft. So one of the data circles which is passed from the aircraft object to the missile object is the location of the aircraft at the moment of launch; another would probably be its attitude, so that the missile will know the direction in which it is headed.

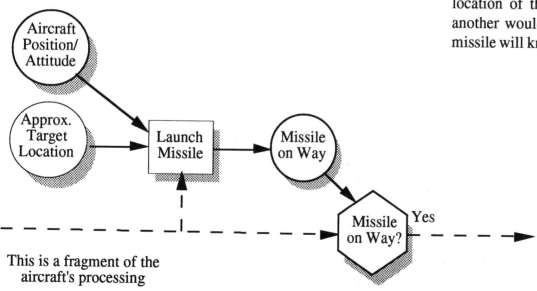

This is a fragment of the aircraft's processing

Software Objects

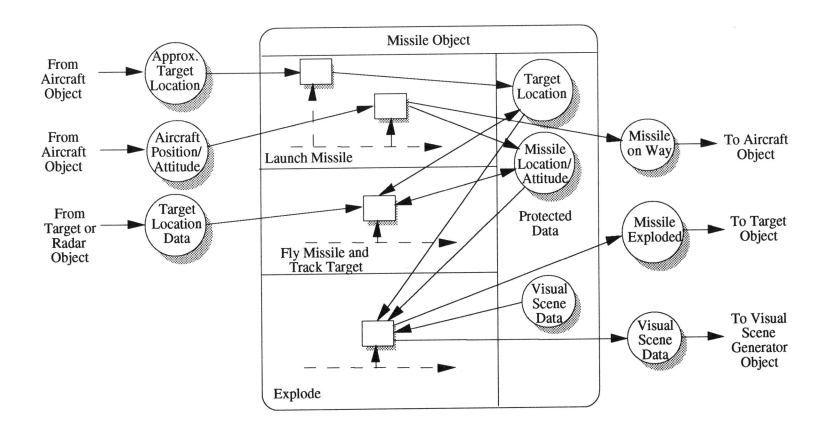

Communicating Concurrent Processes

The processes in parallel control paths may communicate with one another. When this occurs, it can be shown clearly in the notation whether the parallel paths are in separate physical processors or are accomplished by one processor simulating the action of several.

An Automobile Cruise Control

The automobile cruise control system is often used as an example in books on software design methods. Here, it will serve as an example of communicating concurrent processes. There are three actions which are essentially independent: the computation of current speed from the speedometer pulses and clock pulses; the generation of a throttle call signal from the current speed and the desired speed as affected by the "accelerate" button and the "set" button; and the generation of a throttle drive signal from the throttle call, the brake pedal, and the "resume" button.

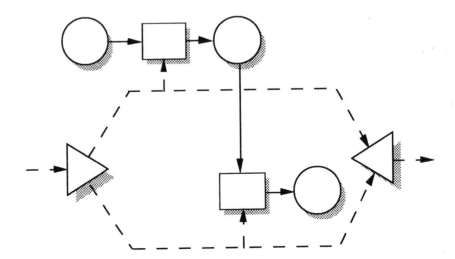

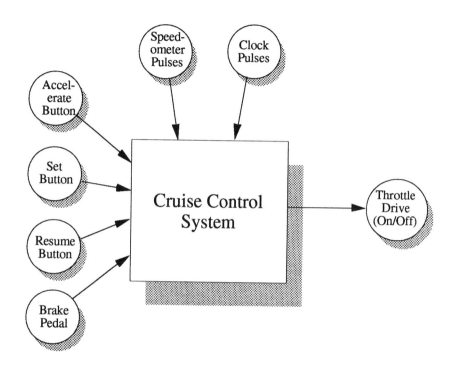

By comparing pulses coming from the speedometer, each representing an increment of distance the automobile has moved, with pulses coming from a clock, each representing a set interval of time, we can calculate the current speed of the automobile. That can be used to set the desired speed (by pressing the Set button) and can be compared with the desired speed to develop a "call for throttle" signal whenever the current speed falls below the desired speed. The call for throttle is not used to drive the throttle directly; it is gated by the Brake Pedal signal and the Resume button, so that the throttle is activated only when it is safe to do so.

Here we show the cruise control system as a set of processes occurring sequentially. If it is executed rapidly enough and often enough, this will work, because everything is computed before it is needed in some other computation. However, for fast response and lower computing requirements, we often want more parallelism; but in any event we will show a parallel solution here in order to investigate the diagramming problems involved.

Communicating Concurrent Processes

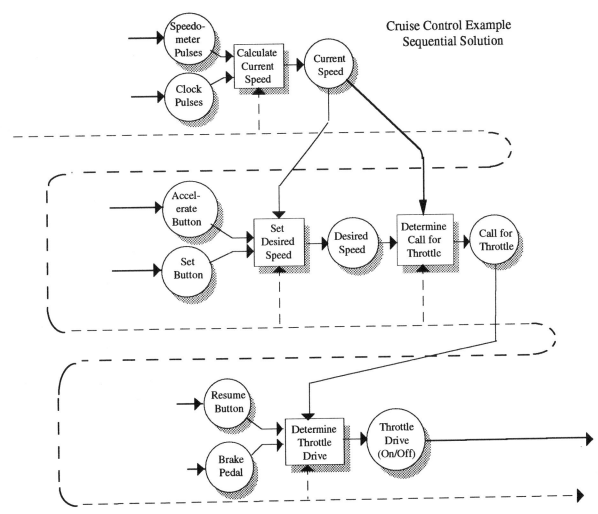

Cruise Control Example
Sequential Solution

The Cruise Control in Parallel Form

However, there are good and bad parallel designs. Here, in our first try, we have broken out three parallel processes: calculating the current speed; setting the desired speed, comparing it with the current speed, and developing the call for the throttle; and accepting the brake/resume controls on the throttle drive. However, the way we've shown them, each can operate once before all three must rendezvous — that is, all must finish their one iteration, and then they will recombine into a single flow path, which returns to the beginning to break out the three parallel paths again. Even though the three paths are caused to stop and wait, correct operation is still not guaranteed — for instance, "Set Desired Speed" may try to access the "Current Speed" while it is being changed by "Calculate Current Speed," resulting in a garbled computation.

We can improve this design somewhat by making the three functions even more independent, so that they don't have to perform a "rendezvous" before they can iterate again.

Communicating Concurrent Processes

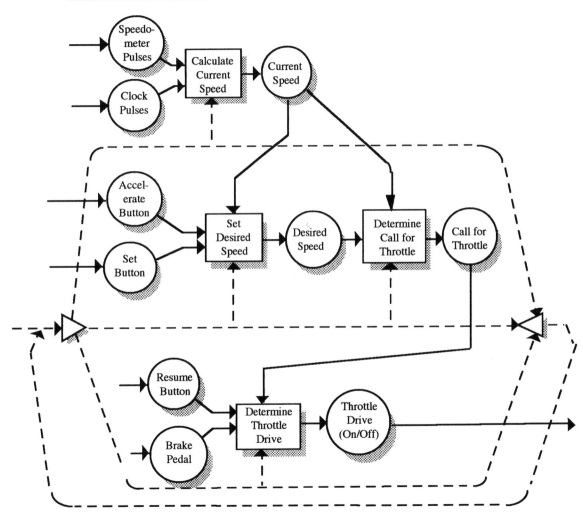

In this design, we have shown the initial single control flow path breaking into three paths which then loop indefinitely (until the system is turned off). That does give the three functions more autonomy.

However, the data integrity problem is still there. In this design, it still isn't guaranteed that "Current Speed" will have been computed by the time "Set Desired Speed" is ready to use it. What we need is an interlock that will prevent the second function from using this quantity until it is ready. The same is true for the "Call for Throttle" signal, which is communicated from the second function to the third.

Communicating Concurrent Processes

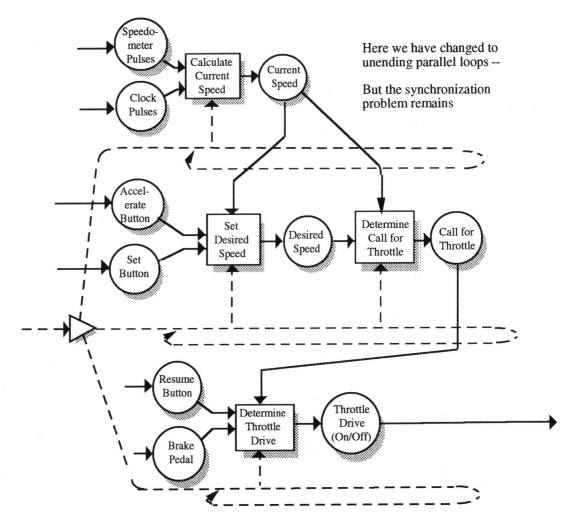

Here we have changed to unending parallel loops --

But the synchronization problem remains

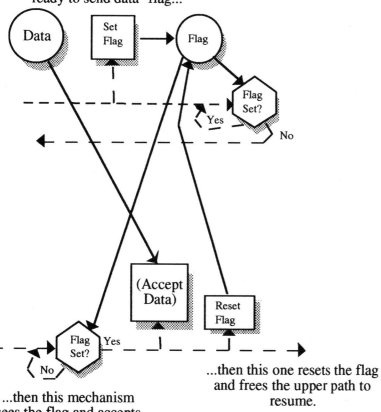

This mechanism sets the "I'm ready to send data" flag...

...then this mechanism sees the flag and accepts the data...

...then this one resets the flag and frees the upper path to resume.

Communication Using Data Flags

Here, we've added a system of flags which cause each function (each parallel control flow path) to halt and wait for the transfer of data from one to the next: "Current Speed" from the first function to the second, and "Call for Throttle" from the second function to the third.

The flags, the processes that set and reset them, and the decision blocks that test them (and then lock or release the interlocks) are all explicitly shown in this notation.

Hiding the Details with a Hierarchy

And while that does show us exactly what is going on, the picture is complicated enough that we would probably like to break it up and show it hierarchically, as we did for the wristwatch example. Each of the major functions can become a single process block, which, if exploded, would show the mechanism we have already explained.

Communicating Concurrent Processes

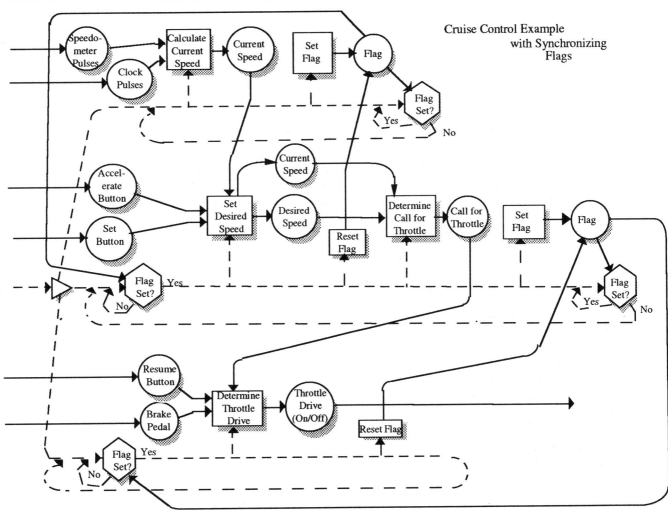

Cruise Control Example with Synchronizing Flags

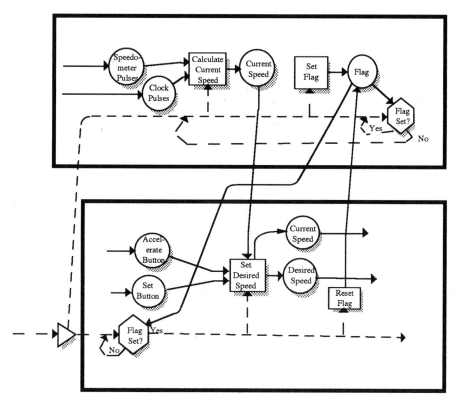

We can begin this process by drawing boxes around the functions we want to become single process boxes. In the case of the first two functions, it will look like this.

However, there is some data that flows from one function to the next, and it will be clearer if that data is shown entering and leaving the processes.

Communicating Concurrent Processes

Here, we have brought that data outside those process blocks: both the quantitative data and the flags that signal the other functions when it is safe to transfer information. We can also bring the unending loop outside so that it is clear at the higher level what is going on. Major inputs and outputs (the speedometer pulses, "set" button, throttle drive, etc.) are also brought out.

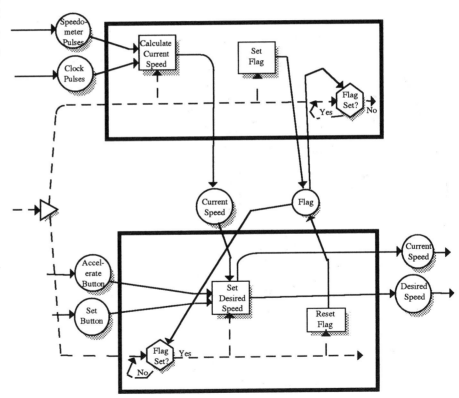

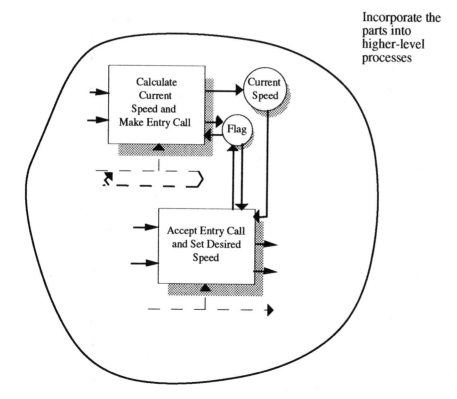

Incorporate the parts into higher-level processes

Next we collapse the boxes into small process boxes with the appropriate text legends in them, as shown at left. When we do that for the third function as well, we can show the whole diagram as we have at right.

We now have a clean representation at a higher level, consistent with the rules of the notation. We can investigate the exploded diagram of each of the major blocks if we want to, and these diagrams will show the internal mechanism of the interlocks.

Actually, the resulting diagram is quite similar to the diagrams one sees of Ada "tasks" communicating with one another, though the flags are typically not shown in those diagrams. At bottom, however, even the Ada mechanisms use data transmissions that actually are flags. They do not show them because it is considered better to hide the details of the implementation. Here, we can hide the details of the rendezvous mechanism we have devised, down within the process blocks; by showing the flags, we form a consistent picture which can be investigated by the user by going down the hierarchy.

Communicating Concurrent Processes

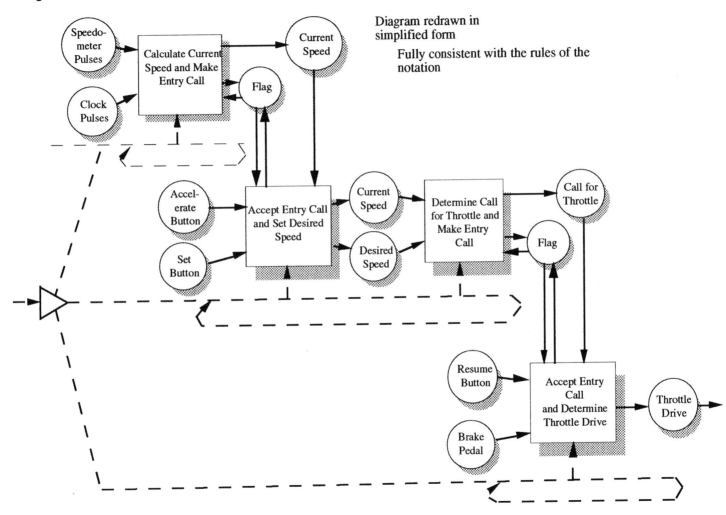

Interrupts

In this notation, control flow is represented by the dotted arrows. In an interrupt, control flow is interrupted anywhere along one of these dotted arrows, or, to be more precise, anywhere within some whole system of dotted arrows, decision blocks, and parallel paths. The computer then goes off to do some predefined processing, and when it's over, sometimes (perhaps most of the time) control will be returned to continue where it left off.

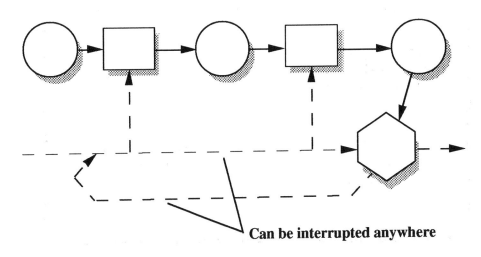

Can be interrupted anywhere

Enclosing Interruptible Software

Since we don't know where the control flow will be when the interrupt comes in, it isn't possible to insert an icon of some sort at some definite position along the dotted arrows. What we have to do is to demarcate the whole range of control flow within which the interrupt can take place; and the most natural way to do that is with one of the enclosures we have used for other purposes. This is just a heavy line surrounding the area that can be interrupted.

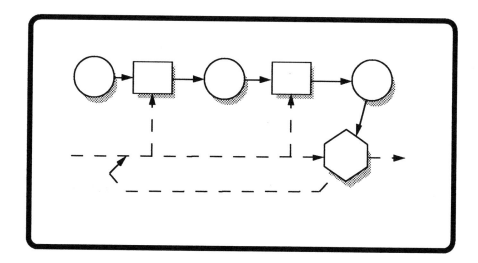

Suspend and Resume Control

To indicate the action of the interrupt, we have defined two special processes (each represented by the usual rectangle): the "Suspend Control" and "Resume Control" processes. We draw solid arrows from those processes to the heavy line that encloses the area which can be interrupted. The suspend and resume control processes operate just like other process blocks. They represent actions that, in the computer, usually take place with the help of special hardware.

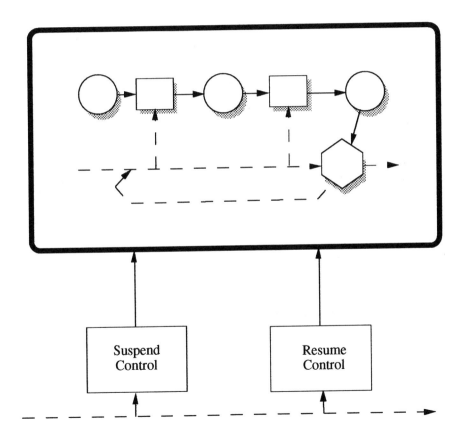

Controlling the Interrupt on a Parallel Path

Normally, interrupts are handled by paths that are separate and parallel to the path that can be interrupted. Typically, special hardware will be monitoring the conditions that can cause an interrupt — for example, an external clock event, an external alarm condition, or an internally-sensed error or special condition. So we can show all that as happening on a parallel path. When the interrupt is triggered, a "Suspend Control" process block interrupts the whole area enclosed in the heavy line. Then some processes are executed to handle the interrupt, after which (sometimes) a "Resume Control" block will be executed to cause the interrupted area to begin executing again. After that, control within the interrupt monitoring path might loop back to watch for the next interrupt, or it might do something else — whatever the program design calls for.

The Interrupt as a Data Signal

Note that the interrupt signal itself is treated as a data signal rather than a control flow action; and, as usual in the notation, a decision block is used to convert a data signal into a control flow action. This is consistent with the rest of the notation, and is also consistent with what is really happening in the computer; interrupt signals really are data until they are used to affect control flow. The interrupts can, of course, also be initiated by the very software that is to be interrupted; in this case a data signal is generated by a process in that software, and then is passed to the wait block in the interrupt control path.

Interrupts

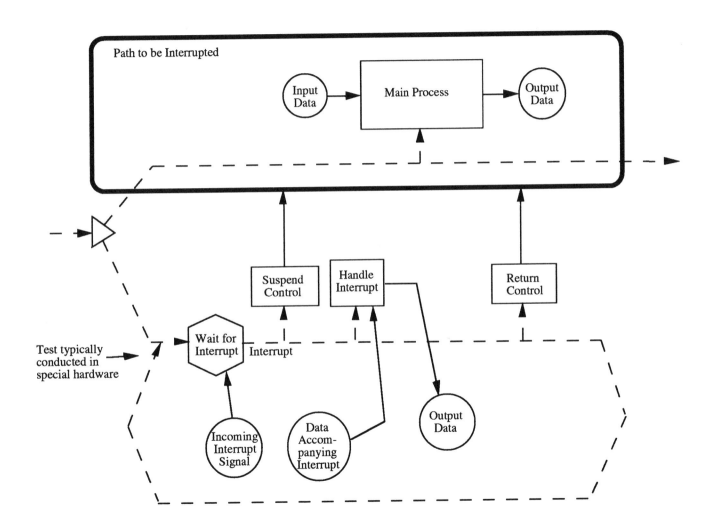

Using the Notation for Software Requirements

This notation can be used to help document the requirements that have been set up for a particular piece of software. Most of the time, what is needed in a particular piece of software is controlled either because it is part of a larger piece of software, or because it is part of a whole process with people, equipment, computers, and software.

By using this notation to draw the larger piece of software, or the larger process, within which the software under consideration will act, we can show the context — including the inputs and outputs.

Functional Requirements

Here, for example, is a very large process — the training of aircraft pilots, shown at an extremely high level. Buried inside the process block "pilot performs task in simulator" is the software for an entire flight training simulator, which might represent a million lines of code, and which assumes the existence of a dozen or more computers, a simulated cockpit, a visual scene generator, hydraulic jacks to impart motion to the cockpit, and much more hardware. This one process block could be expanded out to hundreds of diagrams. The other process block, "Pilot Performs Task in Aircraft," involves other equipment, other computers, and other software, and similarly could be expanded to perhaps thousands of diagrams. But each of these process blocks is a functional requirement; at the level of this diagram, we are saying that there does have to be a function in which the pilot performs tasks in a simulator.

Therefore this diagram becomes a context in which the requirements for the whole flight training simulator can be expressed. Each block shows its own data inputs and outputs, and though they are vague at this level, they again represent requirements.

Explosion of Data Requirements

Of course, each of these data circles really contains within itself a huge number of individual data elements. Those could be broken out hierarchically, level by level. The usual way to do that would be to break them down as the proceses are broken out into lower level diagrams, at the rate of one level of data per level of processes, as was done between diagrams 1.0 and 1.1 in our wristwatch example.

Using the Notation for Software Requirements

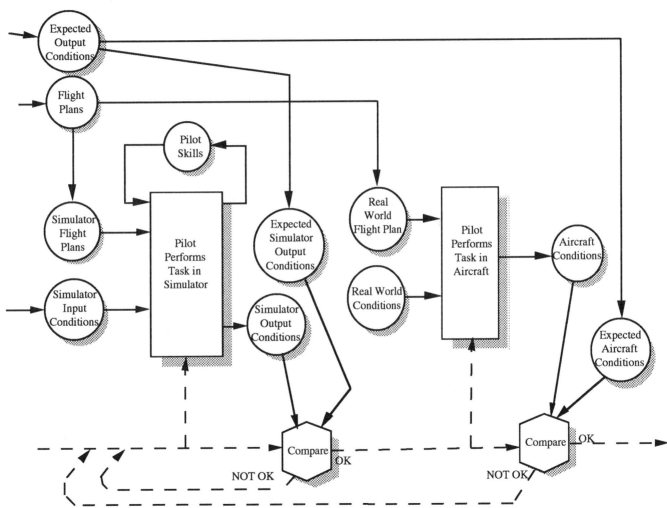

Annotation with Data Attributes

We can also annotate the data circles on the diagram with attributes, for instance accuracy or precision requirements. This is not especially applicable at very high levels, but is relevant after the hierarchical breakdown has reached a level at which the data circles refer to individual data elements. However, even aggregations of data can have certain attributes, such as the revision level of a certain specification from which they should be taken.

Using the Notation for Software Requirements

Annotation with Timing Requirements

Then we can annotate the diagram with time or speed requirements. A requirement for a process block on the diagram to execute in so many milliseconds can be placed directly in or next to the process block.

A requirement for a process block to have executed by a certain time (relative to some zero point) can be written on the dotted control flow arrow just after the dotted arrow stub leading to that process block.

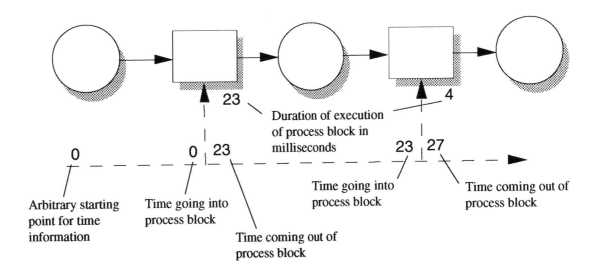

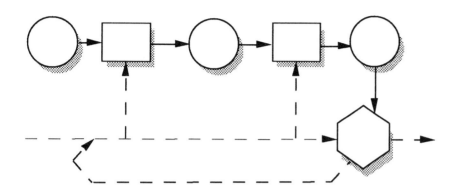

Required Algorithms

If part of the requirements is that a particular algorithm must be used to compute or control something, what is known about that algorithm can be shown on the diagram, by diagramming its known constituent processes, the data elements that are produced as intermediate quantities, and the decisions and control flows that are known to be part of that algorithm. On the other hand, if a particular algorithm is not required – that is, the software designer gets to invent one to suit the purpose – it would be shown just as a single process block with a suitably descriptive legend.

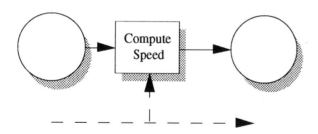

The goal is to include in the diagram as much detail as the requirements state, and no more. Then the diagram, in that form, truly constitutes the requirements for that particular piece of software.

Diagrams at as high a level as the diagram shown at the beginning of this chapter are not often encountered in industry (though they might be at government levels); but on the next level down, diagrams of this sort have been used at CAE-Link Corporation as "operational concept documents" for simulators. These documents are basically high-level expressions of functional requirements.

Rules of the Notation

A **process block** represents executable code. It must be rectangular in shape and bear a legend that identifies the process being executed. A process block must have at least one dotted control flow stub entering, and may have any number. A process block must have at least one data flow input from a data circle, and at least one data flow output to a data circle. A process block may be exploded into a complete diagram in a hierarchical breakdown; in such a case, there must be one control flow arrow entering the diagram and one control flow arrow exiting from the diagram.

A **data circle** represents information that is used in a process block or a decision block, or is produced by a process block. It must be circular or at least rounded in shape, and bear a legend that identifies the information contained in it. A data circle must have at least one data flow arrow entering it or leaving it. In applications where a distinction between electronic and disk memory is essential, a set of concentric circles may be used for disk memory.

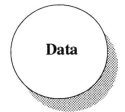

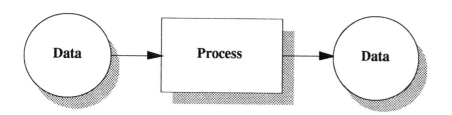

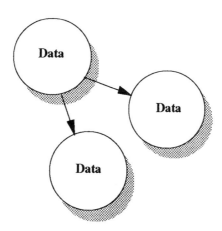

A **data flow arrow** is shown as a solid line with an arrowhead. It represents the flow of data or information and not the flow of control. Data flow arrows may flow from process blocks to data circles, from data circles to process blocks, from data circles to decision blocks, and from data circles to data circles. When data flow arrows flow from a data circle to two or more other data circles, the latter represent component parts of the former. When data flow arrows flow from two or more data circles to a single data circle, the former represent component parts of the latter. When a data flow arrow begins or ends without leaving or entering a data circle, a process block, or a decision block, it represents data coming from, or going to, the parent diagram in the hierarchical breakdown.

Rules of the Notation

A **control flow arrow** is represented as a dotted line with an arrowhead. It models the action of the program counter in a Von Neumann computer, and therefore represents control flow, which proceeds in the direction of the arrowhead. A control flow arrow may have one or more control flow stubs attached to it at right angles, with the arrow on the stub pointing away from the control flow arrow and toward a process block. Process blocks are executed in the order in which their associated control flow stubs are located along the control flow arrow, with those attached to the control flow arrow nearest its arrowhead being executed last. A control flow arrow may end in (point to) a decision block or an opening parallel path triangle, and may originate in a decision block or a closing parallel path triangle. When a control flow arrow originates by itself, that represents the entry of control onto the diagram; there may be only one such entry on a single diagram. When a control flow arrow ends without pointing into a decision block or opening parallel path triangle, that represents the exit of control from the diagram; there may be only one such exit from the diagram if the diagram is an explosion of a process block, while there may be multiple exits if the diagram is an explosion of a decision block. Control flow arrows may merge into other control flow arrows at any point and may loop back to earlier points on the same or other control flow arrows (there is an exception when using **opening parallel path triangles,** which is explained under that heading).

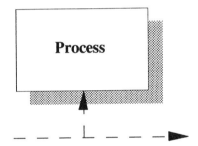

A **control flow stub** is represented as a (typically short) dotted line with an arrowhead, which flows from a point on a control flow arrow to a process block. At the point where it is attached to the control flow arrow, it leaves the control flow arrow at right angles. Process blocks are executed in the order in which their associated control flow stubs are located along the control flow arrow, with those attached to the control flow arrow nearest its arrowhead being executed last.

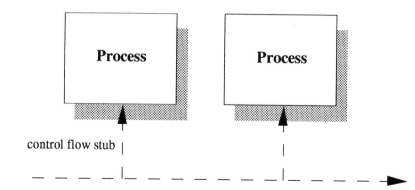

Rules of the Notation

A **decision block** represents a decision, based on data, which selects a control flow path. It must be hexagonal in shape and bear a legend that identifies the question which the decision asks. Legends representing the answers to those questions must be placed on each of the control flow arrows leaving the decision block. At least one data flow arrow coming from a data circle must enter the decision block, and this data must provide the information that is required to make the decision. At least one control flow arrow must enter the decision block, and at least one must exit the decision block. If two or more control flow arrows exit the decision block, it is called a "decision block"; if only one control flow arrow exits the block, it is called a "wait block," and the legend must contain the word "wait." Control flow is trapped within a wait block until the occurrence of the data which will release it. A decision block may be exploded into a complete diagram in a hierarchical breakdown; in such a case, there must be one control flow arrow entering the diagram and the same number of control flow arrows exiting the diagram as exit the decision block which the diagram represents.

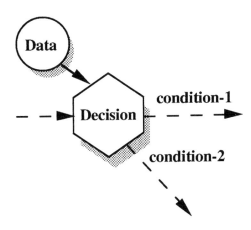

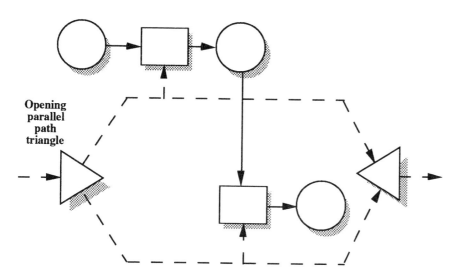

An **opening parallel path triangle** represents the splitting of a control flow path into two or more parallel control flow paths. It is represented as a triangle with one vertical side located where the input control flow arrow enters. The parallel control flow paths leaving the triangle either execute in separate processors or through a valid simulation of execution in separate processors, such as time slicing. An opening parallel path triangle may be used on any control flow path or arrow, even one which is already a parallel path. A parallel path emerging from an opening parallel path triangle may not loop back and merge into the control flow arrow which entered the opening parallel path triangle, without first merging together, in a closing parallel path triangle or a cascade of such triangles, with all other parallel paths spawned from the same opening parallel path triangle.

Rules of the Notation

A **closing parallel path triangle** represents the joining of two or more parallel control flow paths (control flow arrows) into a single control flow arrow. It is represented as a triangle with one vertical side located where the output control flow arrow leaves. Execution on the outgoing control flow arrow does not begin until execution on all incoming control flow arrows has reached the triangle. A closing parallel path triangle may join all parallel paths emerging from one opening parallel path triangle, or any subset thereof, with any paths resulting from the joining of parallel paths emerging from that same opening parallel path triangle.

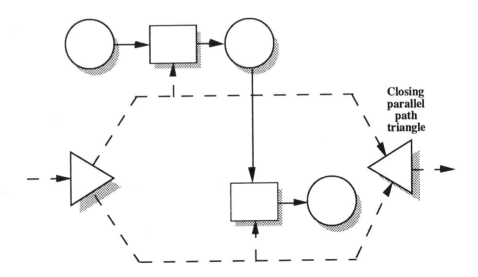

Closing parallel path triangle

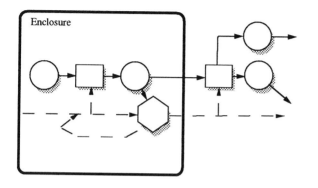

A **heavy line enclosing an area of a diagram** is known as an **enclosure**, and may be added to any diagram for any purpose. There are four uses which are especially recommended:

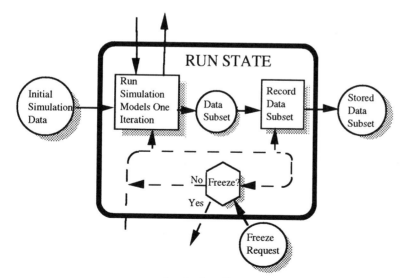

An Active State

State enclosure: a heavy line is drawn around a part of a diagram that represents a software state. Decision blocks either within the states or between the states effect the transitions of control from state to state, which are shown as control flow arrows. States may be active, in which case they contain process blocks, or passive, in which case they do not. Data may be shown either contained within states or as being passed from one state to the next, or may enter or leave the diagram as usual.

Rules of the Notation

Software object enclosure: a heavy line is drawn around a part of a diagram that represents a software object. Within this object, further lines are drawn which separate, first of all, the data which is protected by the object. Other lines are drawn which separate various "operations" of the object, each of which may act on the protected data. The operations are just exploded diagrams of process blocks which may appear on other, higher level diagrams, but are packaged together with other operations which act on the same protected data.

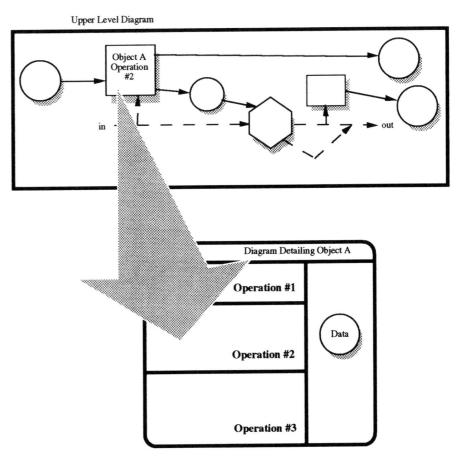

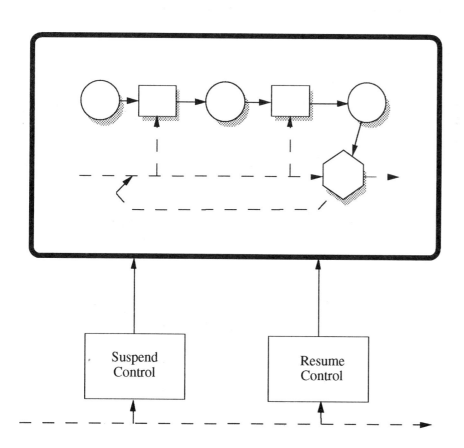

Interrupt enclosure: If a certain area of a diagram represents software which can be interrupted anywhere along its control flow, that area can be placed inside an enclosure which can then be operated on by special processes outside the enclosure, called the "suspend control" and "resume control" processes. These processes, which effect the interrupt, are executed by a control flow path triggered by a decision block; the decision block is fed by the interrupting signal, which appears as a data circle.

Rules of the Notation

Physical CPU enclosure: If a certain area of a diagram represents software which executes in a given physical CPU or some special hardware, that area can be defined with an enclosure.

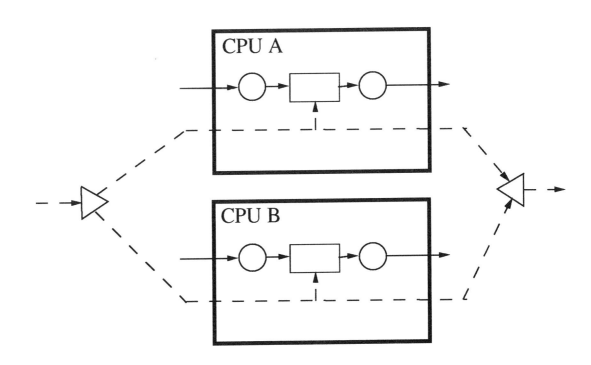

An Agricultural Inspection Device

Consider a food processing plant which receives a certain kind of fruit. It is to be inspected and certain fruit rejected, but we will oversimplify this operation compared to what you would find in a real food processing plant. Let's say that there are three conditions any one of which requires a particular piece of fruit to be rejected:

Fruit which is off-color, that is, when a TV picture is taken of the fruit, the red, green, and blue components of the color at some sampling point are not within a certain percentage of each other.

Fruit which is either too large or too small, again by some set percentage — as measured, let's say, horizontally across its approximate center.

Fruit which is out-of-round, that is, where the maximum vertical and horizontal measurements for a piece of fruit are unequal by some percentage.

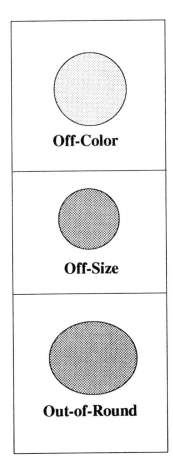

To solve this problem, we will move the fruit, one piece at a time, down a conveyor and past an inspection station. At this station there is a computer-controlled rejector rod, which can push the fruit off the conveyor and onto an entry platform leading to a set of bins. In the set are three bins, one for off-color fruit, one for off-size fruit, and one for out-of-round fruit. Each has an electrically operated, computer-controlled trap door, so that when a piece of fruit is pushed into the entryway, it can be directed to the correct bin.

Also at the inspection station is a TV camera, which views the fruit against a neutral backdrop and passes its picture to the computer to be analyzed. The TV picture is composed of a series of scanlines and some synchronizing signals, which can be separated out.

Finally, there is a fruit detection device, which can signal the computer that a piece of fruit is approaching the inspection station. We assume the existence of devices which will prevent two pieces of fruit from occupying the inspection station at the same time.

An Agricultural Inspection Device

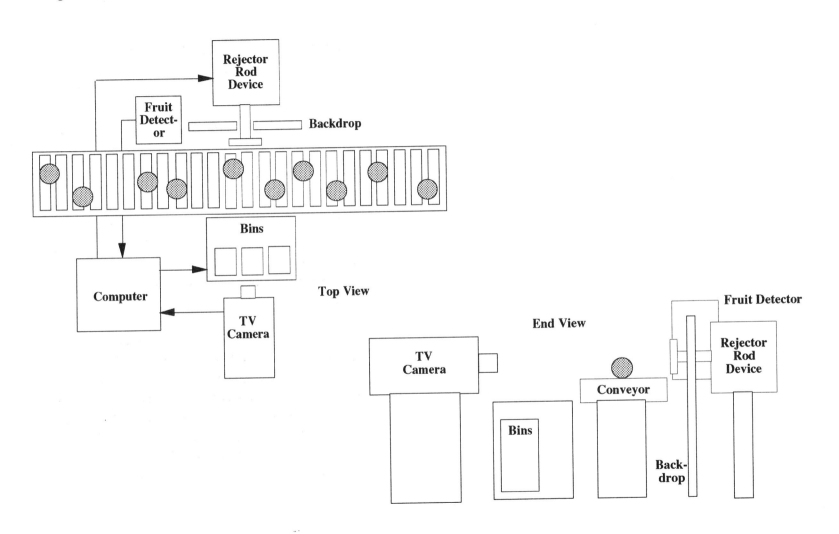

We have devised set of states for the software which analyzes the TV picture and controls the rejector rod and the bin doors. Here the states are represented by enclosures, with the transitions from state to state shown as control flow arrows, and the decision blocks which move the system from state to state shown within the various states.

First there is a test state, in which the whole device can be tested by placing pictures of fruit in front of the TV camera. If there is a problem, an error message is generated, and the system enters a wait state. If all is well, the system makes a transition to the normal run state.

In the run state, the system immediately enters a process block called "Inspect and Sort Fruit." This process waits for a piece of fruit, inspects it and if necessary rejects it, and then ends. This forces the system to drop into the wait state; there, in normal operation, the detection of another piece of fruit will send the system back to the run state. Alternatively, from the wait state an operator can send the system into the test state.

An Agricultural Inspection Device

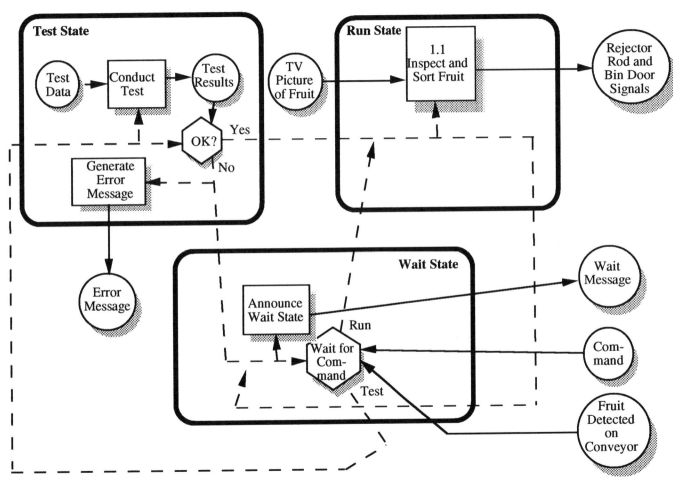

1.0 Fruit Inspection / Rejection States

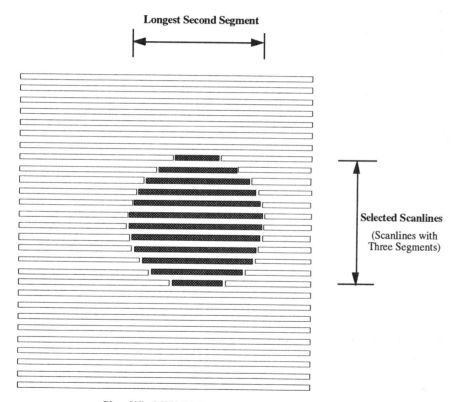

Simplified TV Picture

During the inspection itself, the TV scanlines are analyzed. Because the fruit is viewed against a neutral backdrop, which is chosen to have a color very different from any of the fruit, a single scanline will be either all of the backdrop color (if no fruit intersects the scanline) or divided into segments (if one does). When the fruit has progressed along the conveyor so that it is fully within the picture, the scanlines that intersect the image of the fruit will have three segments, the second of which is the fruit. We can analyze the color of the fruit by isolating this second segment and testing its color.

We can also examine and isolate the group of scanlines that have three segments. The number of these scanlines will give us a measure of the vertical dimension of the fruit. Within these scanlines, we can compare the lengths of the second segments of the various scanlines, and isolate the longest one. This will give us a measure of the horizontal dimension of the fruit. By our ground rules, we can just measure the fruit as it happens to roll by, and ignore the fact that the orientation of the fruit could be hiding a larger or smaller vertical or horizontal dimension.

Of course, by comparing the horizontal and vertical dimensions, we can compute their ratio and therefore determine whether the fruit passes the out-of-round test.

On the next two pages, we show the diagram for this process, "1.1 Inspect and Sort Fruit."

First, we show the TV picture as composed of the scanlines and the synchronizing signals, which we discard. Next, we show a process which selects those scanlines which have three segments. In addition to the actual scanlines, this process also produces a measure of the number of three-segment scanlines (our vertical dimension).

Note that we have not shown, because we have not yet decided, exactly how we are representing each of the scanlines. We could elect to represent them as a long string of individual pixels (picture elements), each described by a red, green, and blue color number. Or we could elect to convert that long description to a short table showing the lengths of the three segments and the red, green, and blue color numbers of the second segment as a whole — perhaps selected from one of its pixels, or computed as an average of all its pixels. At this stage of analysis, we can leave those aspects to be decided at a later time.

Next, we select the longest second segment, and with it, the first and third segments from the same scanline. By comparing these first and third segments to see if they are equal within some tolerance, we can tell if the fruit is centered in the picture (and so centered in front of the rejector rod). If it is not, we go and process the next TV picture.

When the fruit is centered, we can now test for color, size, and out-of-roundness, which we can do in parallel. We compare the length of the longest second segment with the number of three-segment scanlines to determine out-of-roundness; we test the color of the longest second segment; and we test the length of the longest second segment against a constant to see if the fruit is within size limits. When all three tests are done, control passes to a process which uses the outputs of the test to generate rejector rod and bin door signals.

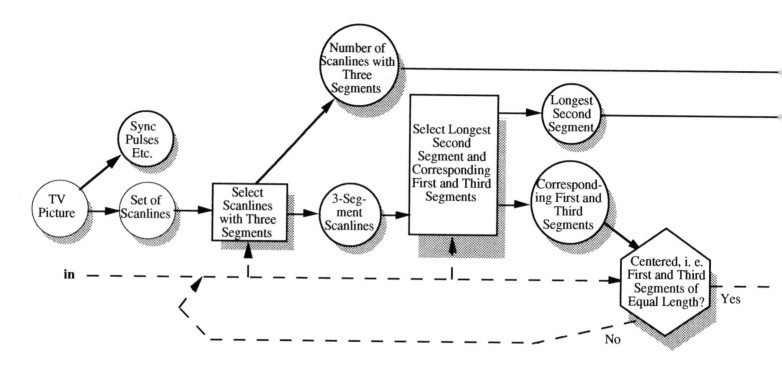

An Agricultural Inspection Device

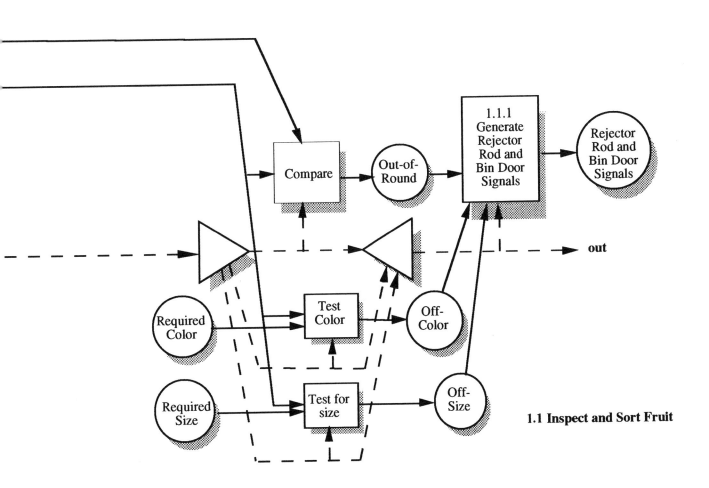

1.1 Inspect and Sort Fruit

Here, we show that process in detail, and can see that there is an order of precedence involved. If a piece of fruit is both off-color and off-size, it will end up in the off-color bin, and so on, according the the order in which the tests are made. We would not have had to design the system that way — we could, for instance, have provided more bins to take care of the fruit that is defective in more than one way — but at least our diagram clearly shows us how the system does work as designed. That might bring up discussion in the design group or between the design group and the using group, which could bring to the surface any problems that exist — and initiate a dialog that would resolve the problems.

An Agricultural Inspection Device

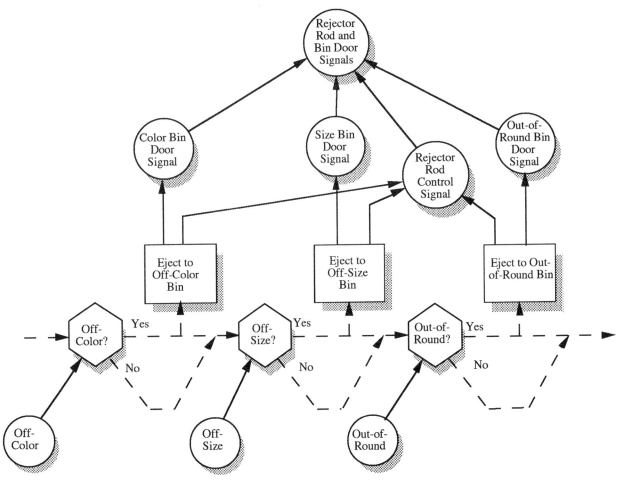

1.1.1 Generate Rejector Rod and Bin Door Signals

A Measurement System

As another example, consider a system which is to measure an electrical voltage and display its value in volts in digital form on a display. This will be an interesting application of the diagramming of interrupts.

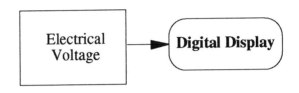

We have elected to solve this problem first of all by feeding the voltage to be measured into a hardware comparator. This is a device that compares two analog voltages, and emits a signal (that can be sent to a computer) when the second voltage equals or exceeds the first. We will arrange to have the second voltage start out at zero and be increased in small steps at closely spaced intervals, perhaps a few milliseconds apart; when we get the comparator signal, we will know that the second voltage is about the same value as the one to be measured.

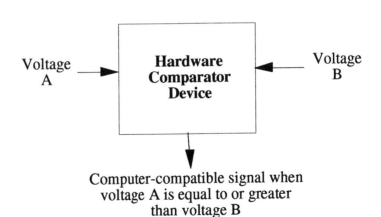

But we will have produced this varying second voltage with a computer by continually incrementing (increasing by one) a digital value inside the computer, then sending this value out to a digital-to-analog converter (another hardware device).

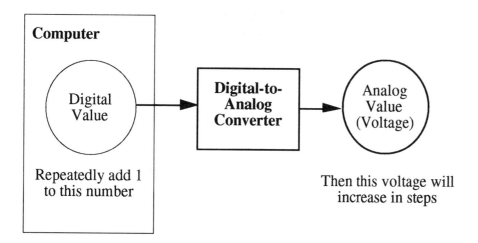

A Measurement System

Because we have produced this value inside the computer, we can stop incrementing it and send it out to the display as soon as the comparator signal is received. This will accomplish our goal of measuring the external voltage and displaying its value in digital form on a display.

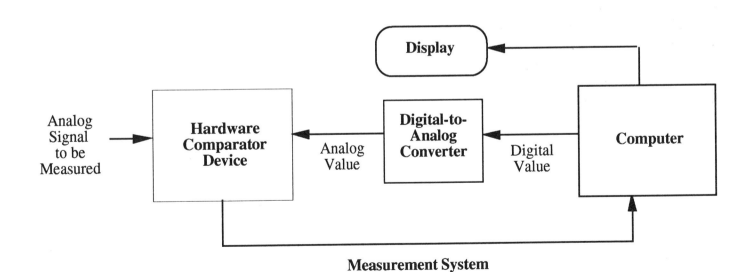

Measurement System

However, we don't know exactly when the comparator signal will come. So we have elected to bring the comparator signal into a second parallel processing path, which will sit and wait for the signal while the first parallel path increments the value that will ultimately be displayed. This waiting-and-looking function is present in most computers as an interrupt chip, which has hardware that can look for an interrupt independently from the main processing unit.

In this diagram, when the system is turned on, the two parallel paths split apart; the upper path first resets the value to zero and then begins an endless loop which continues until the system is turned off. In this loop, it increments the value by one and then sends it out to the digital-to-analog converter.

The lower path waits for the comparator signal. When it comes in, it suspends control on the upper path, wherever the control is. Then it sends the current value to the display, sets the value back to zero, and causes the upper path to resume control where it left off, which will be somewhere in its endless loop. In this case it doesn't matter where in the loop that is; everything will continue to work correctly.

A Measurement System

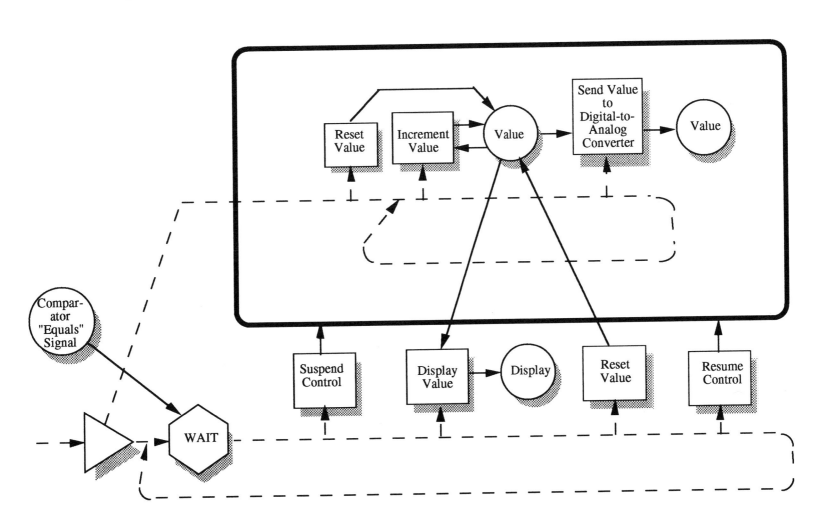

A Simple Camera
Control System

Most of us are familiar with the way a camera works, at least in general outline — which is to say, at the upper levels of the process. So it may be instructive, given that many cameras, even some simple ones, operate through software today, to see how that process is described in our notation.

The camera we are considering is one of these simple ones, at least compared with many modern cameras. It is equipped with a photocell to measure ambient light, an automatically controlled shutter, an automatically controlled iris, an automatic flash, and an autofocus system.

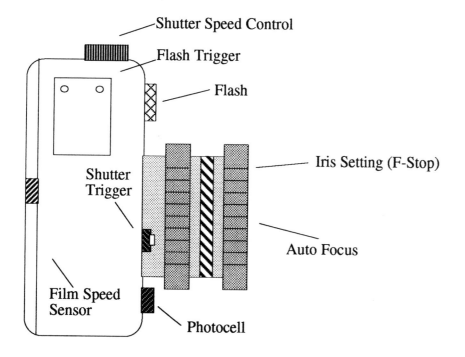

The computer in the camera has the following inputs: the photocell, the autofocus sensor, and a film speed reader which reads the film speed codes on the film can.

The computer's outputs are the iris (setting the F-stop), the shutter speed, the shutter release, the autofocus drive, and the flash (both a flash duration setting and the flash trigger).

From the ambient light level and the film speed, the computer can calculate the amount of light exposure the film needs for this picture (which used to be called the "exposure factor"), and from that it can calculate an appropriate combination of shutter speed and F-stop that will provide that much exposure. In choosing that combination, however, it has to consider depth-of-field, that is, a range of distances from the camera within which the objects in the photograph will be mostly in focus. The smaller the F-stop, the larger that range will be. The camera can use the distance to the subject, obtained from the autofocus system, and a simple rule of thumb to select an F-stop that will provide a reasonable depth of field. How-

A Simple Camera Control System

ever, as camera enthusiasts will know, the determination of a good depth of field cannot really be made without much more knowledge about the subject and the photographer's intentions.

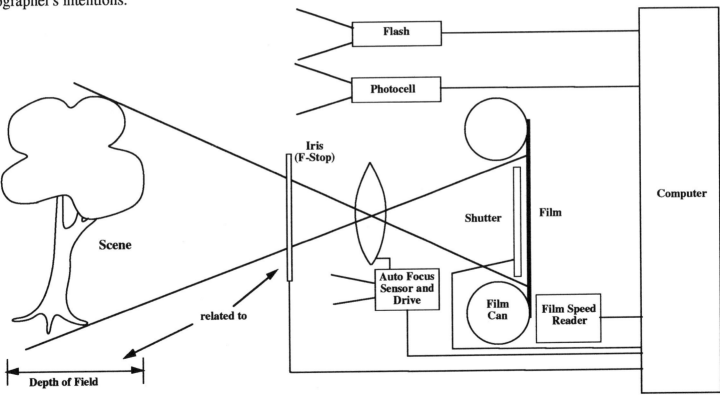

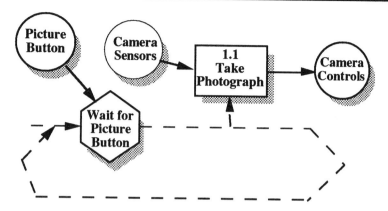

1.0 User-Level Diagram

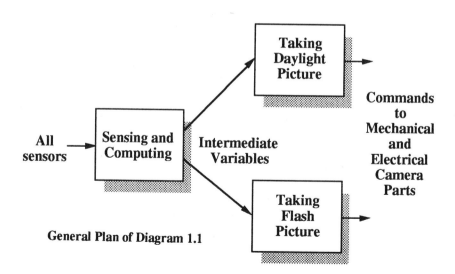

General Plan of Diagram 1.1

Here now are the top two levels of the camera operation in our notation. At left is the highest level, that is, the user level diagram (1.0). At the next level down, 1.1, which contains all that the camera computer itself does, there are three processes: the first does the sensing and computing, and then, as a result of a decision, either the second process takes a daylight picture or the third process takes a flash picture. The processes for these latter two events issue commands that go to the various mechanical and electrical actuators in the camera. All sensors go into the first process, and then only intermediate variables pass from that process to the two later ones.

Note that in this design, there is no provision for a command to wind the film to the next exposure position. That is left for the camera user to do. But there is also no sensor input to inform the computer that this has been done. Consequently there is no protection against double exposures in this camera. But it is clear from these diagrams that this problem exists.

A Simple Camera Control System

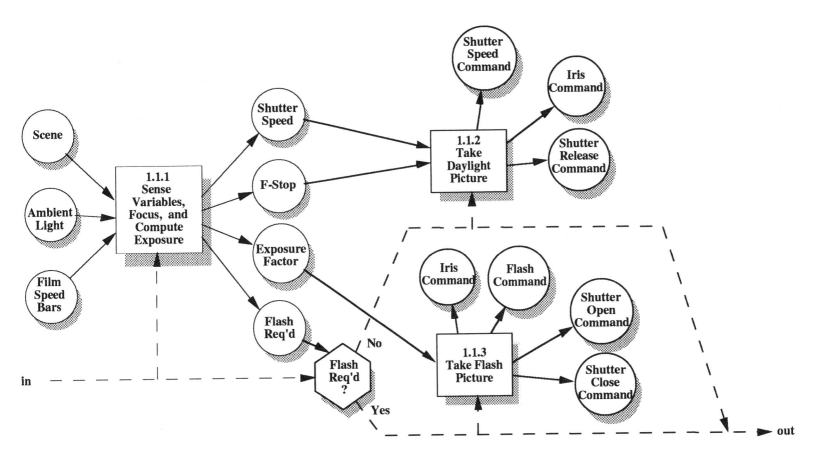

1.1 Take Photograph

The autofocus operation, sensing of ambient light, and sensing of the film speed bars on the film container can all be done in parallel, at least conceptually, and if they are done by separate chips in the camera, would actually be done in parallel. Remember that the meaning of the closing triangle is that all three parallel tracks must finish before control flow can continue beyond that point. In this case, that is necessary before a correct computation can be carried out.

A Simple Camera Control System

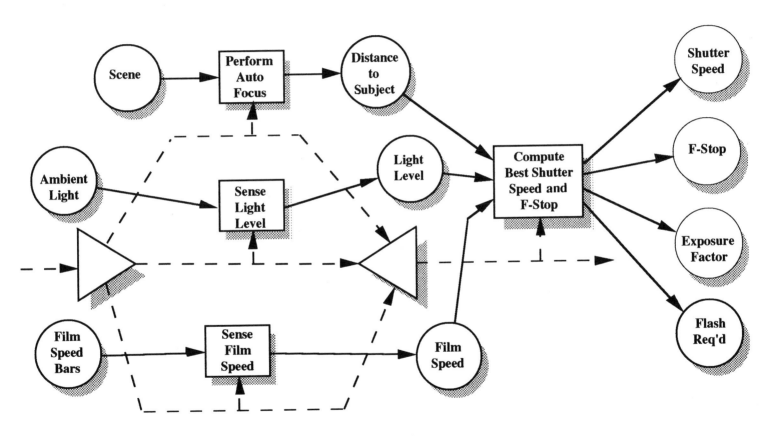

1.1.1 Sense Variables, Focus, and Compute Exposure

In taking a daylight picture, all that is necessary is to set the shutter speed, set the F-stop, and then release the shutter. Of course, both the setting of the shutter speed and the setting of the F-stop, being mechanical operations, take considerable time relative to electronic processes. We assume here that the processes "Set Shutter Speed" and "Set F-Stop" contain programmed delays which account for these times. If they do not, then the shutter release will occur too early, since it will occur as soon as both of the other two processes terminate.

A Simple Camera Control System

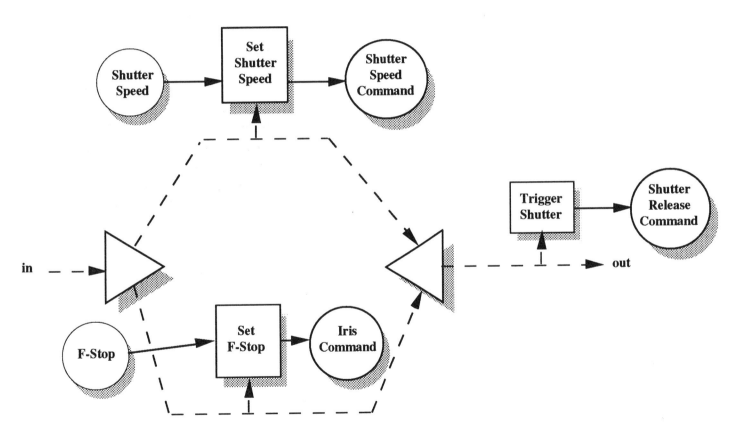

1.1.2 Take Daylight Picture

Flash operations are considerably more complex, although here we use an older system which is nowhere as complex as more modern methods. First, the iris is set to a standard opening which is used with flash. The exposure factor, which is a measure of how much light the picture requires, is used to compute flash duration. However, before the flash can be triggered, the shutter must be opened. This is occurring on a parallel path, so there must be a message passed when the shutter is open. When the flash is triggered, another message is sent back so that the shutter can be closed at the proper time.

A Simple Camera Control System

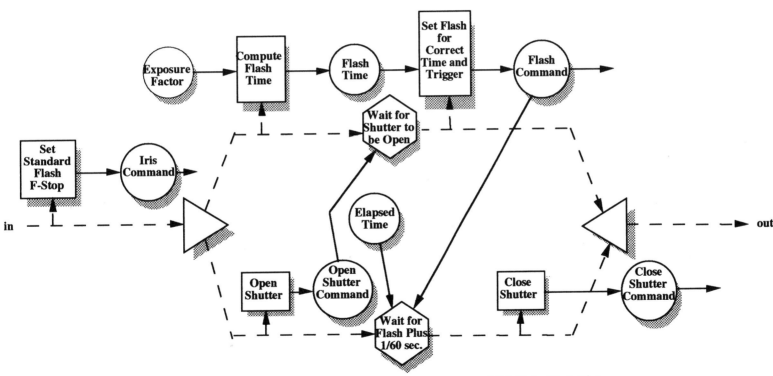

1.1.3 Take Flash Picture

Clarifying Your Drawings: The "Data Bus" Solution

Sometimes a drawing will become quite complex. Then, usually, you would try to break it up hierarchically, collapsing large chunks of the diagram into single process blocks, which are then detailed on their own diagrams. But sometimes there is a single algorithm you would like to keep on one drawing, but you would also like to simplify the diagram without leaving anything out. The following example is one of those cases.

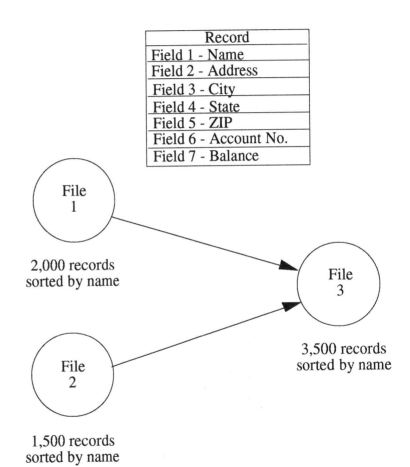

Consider a standard data processing operation, in which we have two sequential files with identical record formats. They are each already sorted on the same field in the record (say, a name field); that is, the records are arranged in order within the files, alphabetically by name. We want to merge these two files into a third file so that all the records from both input files are there, and the output file remains in the same order — alphabetically arranged, by name. Let's also say that we are certain there are no duplicate records.

Because the files are already sorted, a simple comparison method (which predates the computer era) can be used. It involves reading a record from each file, comparing the names on the records to see which should come first on the output file, and then writing out the record with the name that comes first alphabetically. Another record is then read from the file from which that record came, and the process is repeated. If all the records from one file are written, but some remain on the other file; those remaining are then written out.

Clarifying Your Drawings: The "Data Bus" Solution

The software to accomplish this match/merge operation can be partitioned so that separate parts are devoted to the management of each of the three files. These would become software objects, since they each protect their own file. First, an object is devoted to each of the input files; if the main algorithm wants to read a record from one of those files, it must invoke the appropriate operation of one of those objects. These "reader" objects produce the individual records, and also a flag when the file is exhausted.

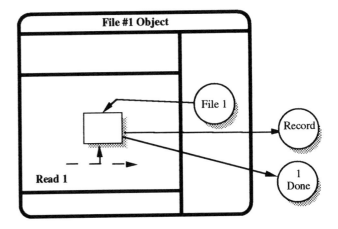

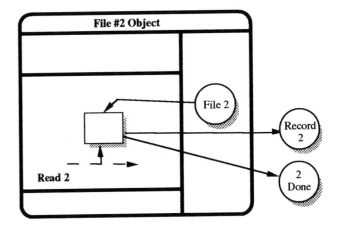

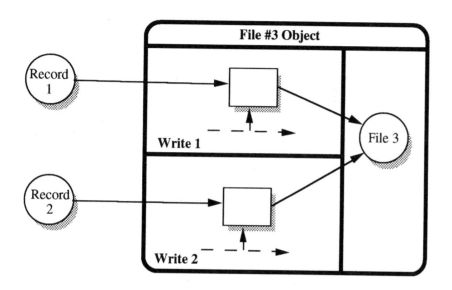

Next, we devise a writer object to take care of the output file (File 3); it will have two operations, one to write a record from File 1 and the other to write a record from File 2.

All three of these objects could be extended to show how they deal with calls to a specific operating system, how buffers are handled if that must be done by the application program, and the handling of various hardware error conditions.

Finally, there will be a controller object which organizes and coordinates the whole match/mege. This is the algorithm we didn't want to break up. In order to make it easier to comprehend, we have provided a set of parallel solid data flow arrows at the top, in a manner quite similar to the data bus structures we see in electronic hardware diagrams. Each of these terminates in a data circle, and also connect the circle to various process boxes. The rules for this diagram are exactly the same as for all our other diagrams, but arranging the diagram this way makes it much cleaner than it would be otherwise, given that each input and output circle interacts with several process blocks. Note also that there are multiple instances of process blocks which refer to the same operation of the same object.

Clarifying Your Drawings: The "Data Bus" Solution

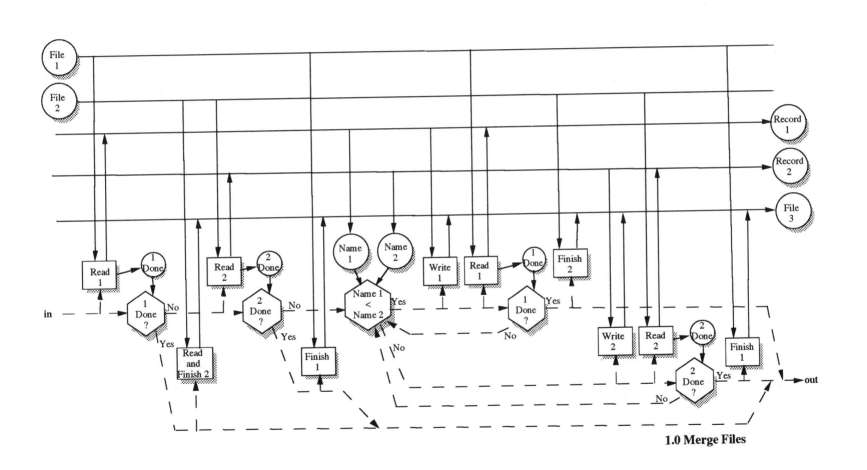

1.0 Merge Files

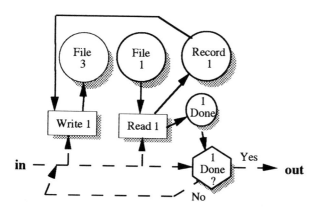

Finish 1

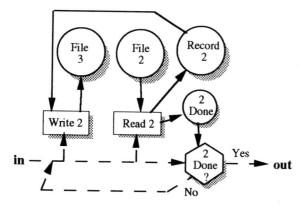

Finish 2

This algorithm refers to three more process blocks: "Finish 1," "Finish 2," and "Read and Finish 2".

Finish 1 is used when all the records from File 2 have been transferred to File 3, and now we need to transfer the remaining File 1 records. Finish 2 does exactly the same thing for File 2 when File 1 has been completely transferred, but File 2 records remain.

Clarifying Your Drawings: The "Data Bus" Solution

"Read and Finish 2" must be used when File 1 is discovered to be empty when the attempt is made to read its first record. A corresponding piece of code is not required to read and finish File 1, since File 1 is read first in the main algorithm.

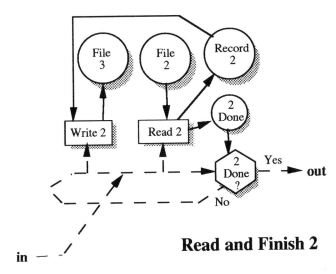

Read and Finish 2

Choosing a CASE Tool

A graphical notation for software is much easier to use if a computer-based tool is employed. Of course, it is possible just to use a graphics tool of the "draw" type, such as MacDraw for the Macintosh. The drawings in this book were done with MacDraw II Version 1.1, because they are then especially suitable for publication.

For intensive software design work, however, tools with specialized capabilities are available. Usually these are called CASE tools, for Computer-Aided Software Engineering. Many of these tools are based on particular software notations and can't be used with any other notation; of these, the oldest ones are built around flow charts, and the ones that date from the late 1970's or early 1980's have usually been built around data flow diagrams.

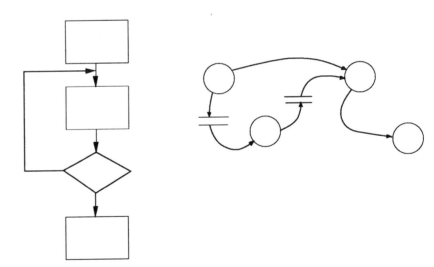

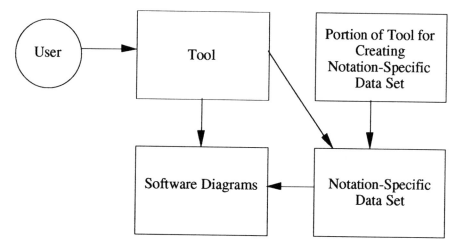

Now, however, a number of tools exist that are fully programmable — that is, they can be adapted to any notation, and hence can be adapted to the one described in this book. To adapt one of these tools to a notation, a data set is created that describes the icons and rules of the notation, usually in a format peculiar to the tool. Once that's done, however, it doesn't have to be repeated, and the data set can be put up for sale — although the user also has to buy a software license for the tool itself. Creation of these data sets is usually not very difficult, typically requiring not more than a few person-weeks.

These tools are not, however, all created equal. They come with different capabilities, not all of which can be programmed by the user if they are not there to begin with. So we provide here a discussion of the capabilities to look for when choosing one of these tools for the notation described in this book.

Choosing a CASE Tool

Probably the most important capability is to be able to move sections of a drawing around while keeping the connections — the data flow arrows and control flow arrows — intact. Then you can insert new material into the diagram in the space you have opened up. What you want to be able to do is to identify a particular part of the drawing and then move that part, as a whole piece, somewhere else on the drawing.

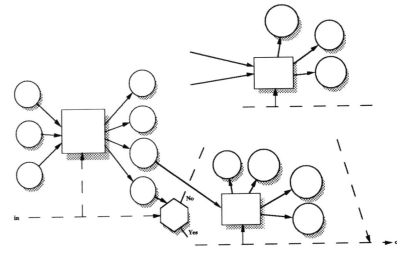

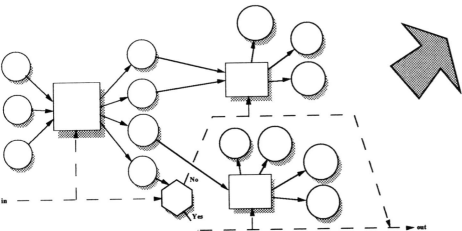

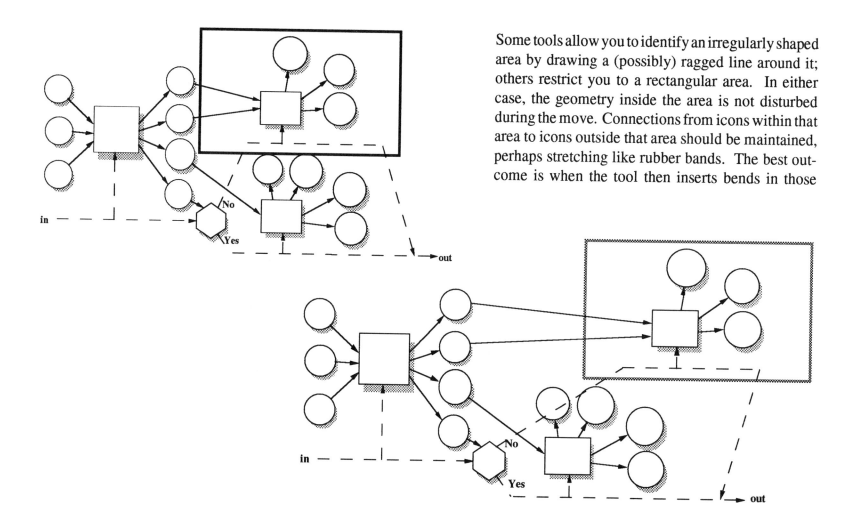

Some tools allow you to identify an irregularly shaped area by drawing a (possibly) ragged line around it; others restrict you to a rectangular area. In either case, the geometry inside the area is not disturbed during the move. Connections from icons within that area to icons outside that area should be maintained, perhaps stretching like rubber bands. The best outcome is when the tool then inserts bends in those

Choosing a CASE Tool

connecting lines where necessary to make the flow as graceful as possible — and at least to avoid having the connecting lines pass through other icons. However, most tools are not yet capable of that, but you should look for an easy ability to insert those bends yourself, after the move.

Sometimes it improves the readability of a diagram considerably to have smooth bends in connecting arrows as opposed to sharp ones, although this diagram here is simple enough so that such a refinement doesn't matter.

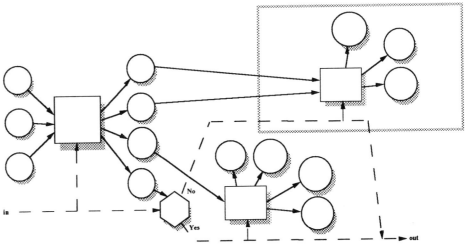

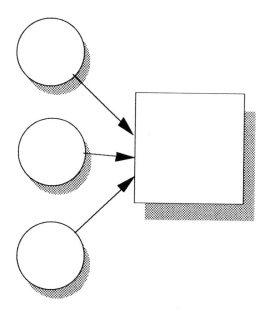

These tools usually allow you to design icons with specific points to which the connecting lines will attach, and sometimes that's exactly what you want. Our notation, however, allows you more freedom than that; for instance, a data flow arrow can enter or exit a data circle at any point on its periphery. I usually draw diagrams so that the data flow arrows enter or exit the circles in such a way that they would pass through the center of the circle if extended. That is an excellent capability to have in a tool, and some tools do provide it.

Choosing a CASE Tool

When you connect to a certain place on an icon, the tool should automatically select the correct type of connecting arrow for that point—for instance, in our notation, either a data flow arrow or a control flow arrow.

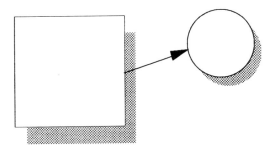

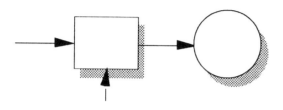

An Often-Used Fragment

When you are creating new diagrams or adding significant amounts to an existing diagram, it is very helpful to be able to construct the diagrams from fragments containing several icons and connecting arrows. The tool should provide you a way to construct these often-used fragments and to store them for easy use later. While this is quite easy to do with the "draw" programs like MacDraw II, it is sometimes not provided with more complex tools.

An important capability is to be able to "click" on an icon and then have a window appear on the screen containing data about that icon. For instance, you should be able to click on a particular data circle and get a window into which you can put the attributes of that data — perhaps even code declarations — and then see them any time it's necessary. When the data circle represents an aggregate of individual data items, you should be able to see that.

We might wish for automatic code generation when the drawing reaches a sufficiently low level. This can't be done with many notations, but it can be with ours, since the information you need is there — if the necessary details about the data items have been entered into the windows mentioned above.

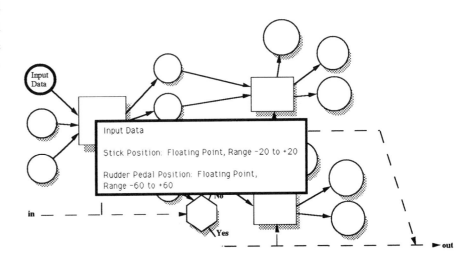

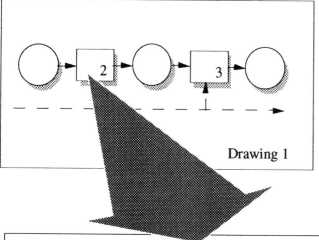

Drawing 1

In a similar way, you should be able to click on a process box and have the display change to the drawing which details that process box ("explodes" it). Since decision blocks can be exploded also, you should be able to do this with decisions as well. You should be able to go down any number of levels, and then work your way back up along the same path.

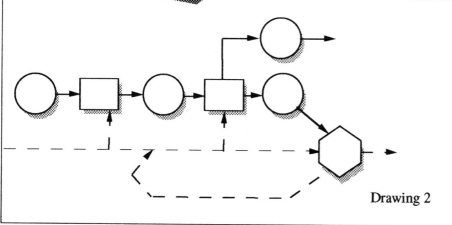

Drawing 2

Choosing a CASE Tool

Most of the more sophisticated tools provide for "data balancing," which means (in our notation) that if you explode a process block, the data circles which are shown entering and leaving that process block must match exactly the circles which enter and leave the exploded diagram. The check on this should be automatic, with error messages provided if the drawing doesn't "balance." Some tools automatically bring the input and output data circles from a process block down to its exploded drawing when you first begin to draw that exploded drawing.

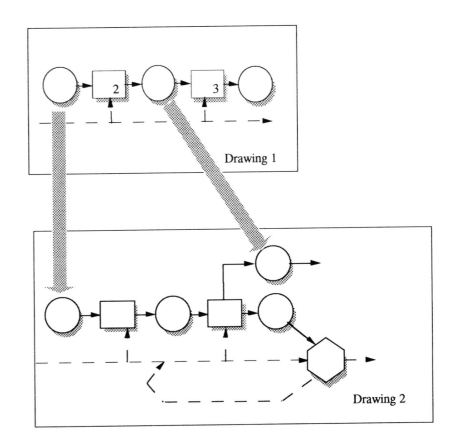

Drawing 1

Drawing 2

This automatic bring-down of data can include other things, such as the notes on precision, accuracy, and timing that constitute your requirements annotations.

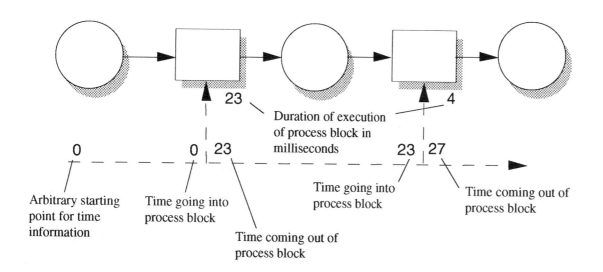

Among the most useful of CASE tools is the "reverse engineering" tool, which can take code and (in this case) produce the diagrams from it. In most cases this can't be done perfectly because the hierarchical grouping of functions, and the names given to those functions, may not have been done in the code. Some tools of this type depend upon the insertion, in the code, of comments that control these aspects. Usually the comments contain special codes that the diagramming tool can recognize.

These tools are especially useful for converting designs written in pseudo-code (or "design languages") into diagrams, and for converting code written for rapid prototypes into diagrams, which can then be used for reviews of the overall design strategy.

A Closing Note

We hope that you will find this notation as useful for software design as we believe it to be. We hope especially that you will use it for discussions about your software, because we believe it will give you the clearest and most precise forum for such discussion that could be provided.

As it turns out, with only one added icon this notation can be used for processes that include both people and assorted tools. Thus it can be used for factory flow planning, office work planning, and many other uses. To implement that, we have included a dotted circle, which is used just like a data circle would be used, to indicate people or tools; the dotted circle is connected to a process box by a solid arrow, and indicates that the person or tool is to be used in that process block.

We would be very pleased to hear from any user who utilizes our notation.

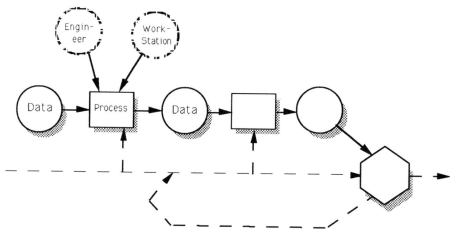

References

1. Miller, George A., "The Magical Number Seven, Plus or Minus Two: Some Limits on Our Capability for Processing Information," *The Psychological Review*, Vol. 63, No. 2, March 1956, pp. 81-97.

2. Yourdon, Edward and Larry L. Constantine, *Structured Design: Fundamentals of a Discipline of Computer Program and Systems Design*, Englewood Cliffs, NJ: Prentice-Hall, 1975.

3. DeMarco, Thomas, *Structured Analysis and System Specification*, Englewood Cliffs, NJ: Prentice-Hall, 1978.

4. DeKleer, Johan and John Seely Brown, "Mental Models of Physical Mechanisms and Their Acquisition," in *Cognitive Skills and Their Acquisition*, edited by John R. Anderson; Hillsdale, NJ: Lawrence Erlbaum Associates, 1981.

5. Gentner, Dedre and Albert L. Stevens, *Mental Models*, Hillsdale, NJ: Lawrence Erlbaum Associates, 1983.

6. Ward, Paul T. and Stephen J. Mellor, *Structured Development of Real-Time Systems*, Vols. 1-3, Englewood Cliffs, NJ: Yourdon Press, 1985.

7. Hatley, Derek J. and Imtiaz A. Pirbhai, *Strategies for Real-Time System Specification*, New York: Dorset House, 1987.

8. Woodard, Pamela S. and William F. Parrish, Jr., "Feasibility of a Graphical Design for an Ada Software Development," Proceedings of the Industry/ Interservice Training System Conference, 1988, pp. 35-40.

9. Meyer, Bertrand, *Object-Oriented Software Construction*, Englewood Cliffs, NJ: Prentice-Hall, 1988

10. Yourdon, Edward, *Modern Structured Analysis*, Englewood Cliffs, NJ: Prentice-Hall, 1989

11. Coad, Peter and Edward Yourdon, *Object-Oriented Analysis*, 2nd Edition, Englewood Cliffs, NJ: Yourdon Press, 1991.

12. Rumbaugh, James, Michael Blaha, William Premerlani, Frederick Eddy, and William Lorensen, *Object-Oriented Modeling and Design*, Englewood Cliffs, NJ: Prentice-Hall, 1991.

Index

Active software states, 77
Ada tasks, 108
Advantages
    of control flow arrows, 20
    of the notation, 15
Aggregations, of data, 39
Agricultural inspection device, 137-147
Alarm/display control, in watch example, 53
Alarms?, in watch example, 67
Alarm set, in watch example, 57
Algorithms, as requirements, 122-123
Alphanumeric design representation, 2
Annotation, with data attributes, 120
Annotation, with timing requirements, 121
Arrows
    connecting, in CASE tools, 181
    rules for, 126-128
Aspects of software, 3, 6, 8
Automobile cruise control example, 97-109
Balancing
    of data in CASE tools, 185
    of software diagrams, 45
Behavior, of software, 2
Block diagram, in measurement example, 151
Bus, data, diagram, 171
Calls, to an operation of an object, 82
Camera control system example, 155-165
CASE tools, vi
    choosing, 175-185
    programmable, 176
Changes, in software, 87-89
Chime set, in watch example, 56
Chrono time display/control, in watch example, 61
Clarification of drawings, 167
Closing note, 188
Code generation, in CASE tools, 183
Communicating concurrent processes, 97-109
Communication
    among software objects, 91-95
    by messages, 47
    using data flags, 104
Comparator, in measurement example, 149
Compile-time inheritance, 86
Computer-aided software engineering,
    tools for, 175

Index

Connecting arrows, in CASE tools, 181
Connection points, in icons, 180
Control flow, iii, 6, 9
   action on processes, 19
   arrows, 19
     rules for, 127
   diagrams, 7
   parallel, in camera example, 160-161
   stubs, 19
     rules for, 128
Controller object, 170-171
Control system, in camera example, 155-165
CPU enclosures, 80
   rules for, 135
Cruise control
   example, 97-109
   parallel form, 100-109
   sequential form, 98-99
Data attributes, annotation with, 120
Data balancing, in CASE tools, 185
Data bus diagram, 171
Data bus solution, 167-173
Data circles, 18
   rules for, 125

Data flags, 104
Data flow, iv, 9
   analysis, iii
   arrows, 18
     rules for, 126
   diagrams, 1, 6
   in hierarchical diagrams, 37
Data processing example, 167-173
Data protection, iii
Data requirements, 118-120
Data, action on decision blocks, 21
Data, hierarchies of, 39
Debugging, of software designs, 33
Decision block hierarchy, in watch example, 67
Decision blocks, 9, 21-24
   and state transitions, 76
   rules for, 129
   with one output, 23
Decisions, iv
   hierarchies of, 38
Design example, wristwatch, 41-73
Design, hierarchical, 33
Design languages, 3
Design representations, 2

Design reviews, 33
  in sorter example, 146
Design-time inheritance, 86
Diagrams
  interconnections between, 31-38
  large, 12
Digital-to-analog converter, 150
Discussion, of software, 33
Display, in wristwatch example, 41-42
Do-Until, 27
Do-While, 26
Draw type tools, 175
Dual time display/control, in watch example, 57
Dual time set, in watch example, 57
Elaboration-time inheritance, 86
Encapsulation, iii
Encapsulation, in control of changes, 87
Enclosures, v, 14, 75-80
  in interrupts, 112
  for physical CPU's, 80
  rules for, 132-135
  of software objects, 81
  for software states, 76-79
Execution, of processes, 19

Explosion, of processes in CASE tools, 184
File processing, 167-173
Flat connections, 31
Flight simulators, v
Flight training simulation, use of objects in, 88-95
Flow charts, 1, 6
  similarity of control flow to, 22
Fragments, of drawings in CASE tools, 182
Fruit sorter, 137-147
Functional requirements, 118
Generation, of code in CASE tools, 183
Graphical design representation, 2
Graphical notations, 3
Hexagonal decision blocks, rules for, 129
Hiding details, with hierarchy, 104-109
Hierarchical design, 33
Hierarchical diagrams, 32-39
  in watch example, 55
Hierarchical organization, iv
Hierarchical structure, of software, 7
Hierarchies
  of data, 39
  data movement in, 48
  of decisions, 38

Index

of maps, 11
of processes, 36
of software, 11, 13
used to hide details, 104-109
Hierarchy charts, 7
High-level diagrams, 119
If-Then-Else, 25
Inheritance, in software objects, 85-86
Insertion, of icons in CASE tools, 177-179
Interactions, 5
of software aspects, iv
Interlocks, 104-109
Interrupts, 111-115
control of on a parallel path, 114
as data signals, 114
enclosure, rules for, 134
in measurement example, 152-153
Large diagrams, 12
versus small diagrams, in watch example, 55
Light control, in watch example, 73
Localization, of changes to software, 87-89
Looping, of control flow arrows, 22, 23
Map analogy, 4-5, 11
Measurement system example, 149-153

Measuring time, in watch example, 69
Memory, 3
Mental simulation, iii, iv, 2, 3, 8
Merging of records, 167-173
Merging, of control flow arrows, 22
Message
passing, among software objects, 84
sending, in watch example, 55
Messages
among software objects, example, 91-95
communication by, 47
Methods, software, vi
Mini-specifications, 3
Modes, in watch example, 42, 45-47
Moving, of icons in CASE tools, 177-179
Nesting, of enclosures, 75
Normal Time Display/Control, in watch example, 49
Normal time checks, in watch example, 71
Normal Time Set, in watch example, 50
Notation
advantages of, 15
use in design, 33
used as process diagram, 189
Notations, of software, iii, 1, 3, 8

Object
  classes, 1
  controller, 170-171
  enclosures, 81
    rules for, 133
Objects, software, iii, v, 1
  see also software objects
  in control of software changes, 87-89
  in data processing example, 169-170
  in flight training simulation, 88-95
  in industrial software, 87
  inheritance in, 85-86
  message passing among, 84
  operations of, 82
Operational concept documents, 123
Operations, of software objects, 82
Parallel control flow
  in camera example, 160-161
  example, 97-109
  paths, 10, 28-29
  symbols, 28-29
  triangles, rules for, 130-131
  in watch example, 43-44
Partitioning, of software, iii, 1

Passive software states, 77
People, processes including, 189
Physical CPU enclosure, 80
  rules for, 135
Process blocks, 18
  rules for, 125
Process diagrams, use of notation for, 189
Processes, iv, 9
  including people, 189
  hierarchies of, 36
Programmable CASE tools, 176
Pseudo-code, 2
Real-world correspondence, 1
  in objects, 90-95
Record merge example, 167-173
Reminder control, in watch example, 53
Rendezvous
  of parallel paths, 10
  rules for, 131
Required algorithms, 122-123
Requirements
  documentation with the notation, 117-123
  for data, 118-120
  functional, 118

Index

for timing, 121
Resume control, in interrupts, 113
   in measurement example, 152-153
Reverse engineering, 41
Review, of software design, in sorter example, 146
Ring alarms, in watch example, 65
Rubber-banding, in CASE tools, 177-179
Rules of the notation, 125-135
Run-time inheritance, 86
Scanlines, in sorter example, 142
Segmentations, of software, v
Separation through enclosures, 75-80
Set timer time setting, in watch example, 61
Simulation, mental, iii, iv, 2, 3, 8
Software
   aspects, iv, 3, 6, 8
   behavior, 1
   changes, 1
      control of, 87-89
   interfaces, 1
   methods, vi
   notations, iii, 1, 8
   objects, iii, 1, 81-95
      controller, 170-171
      in data processing example, 169-170
      enclosure, rules for, 133
      in flight training simulation, 88-95
      in industrial software, 87
      inheritance in, 85-86
      message passing among, 84
      operations of, 82
Software requirements, documenting, 117-123
Software states, 76
   enclosure, rules for, 132
   in sorter example, 140-141
Software tools, vi
State
   enclosure, rules for, 132
   transition networks, 3, 76
States
   active and passive, 77
   in agricultural sorter example, 140-141
   in flight training simulator example, 78-79
   of software, 76
Structured constructs, 25-27
Structuring diagrams, 34-35
Stubs, control flow, rules for, 128
Suspend control, in interrupts, 113

in measurement example, 152-153
Synchronization
  with data flags, 104-109
  problems, in parallel processes, 100-103
Tasks, as in the Ada language, 108
Time Display and Control, in watch example, 45-47
Timer time display/control, in watch example, 61
Timing requirements, annotation with, 121
Tools, software, vi
Top-level diagram, in wristwatch example, 43
Tradeoffs, in selecting objects, 92

Transitions, of states, in sorter example, 140-141
Triangles, parallel path, rules for, 130-131
TV scanlines, in sorter example, 142
Use, of notation for process diagrams, 189
Visualization, 1
Wait blocks, 23-24
  example of, 50
  in measurement example, 153
  rules for, 129
Windows, of data about icons, 183
Wristwatch example, 41-73